Renewable Energy Law and Development

Renewable Energy Law and Development

Case Study Analysis

Richard L. Ottinger
Pace University School of Law, USA

Edward Elgar
Cheltenham, UK • Northampton, MA, USA

Published by
Edward Elgar Publishing Limited
The Lypiatts
15 Lansdown Road
Cheltenham
Glos GL50 2JA
UK

Edward Elgar Publishing, Inc.
William Pratt House
9 Dewey Court
Northampton
Massachusetts 01060
USA

A catalogue record for this book
is available from the British Library

Library of Congress Control Number: 2013933499

This book is available electronically in the ElgarOnline.com
Law Subject Collection, E-ISBN 978 1 78254 664 1

ISBN 978 1 78254 663 4

Typeset by Servis Filmsetting Ltd, Stockport, Cheshire
Printed and bound in Great Britain by T.J. International Ltd, Padstow

Contents

Contributors

Sayan S. Das (India) is an LL.M. graduate of Pace Law School. He received his law degree from Pune University in Pune, India and was a Research Assistant for Dean Ottinger.

Lia Helena M.L. Demange (Brazil) received an LL.M. degree from Pace Law School and is a law graduate of the University of São Paulo. She was a Research Assistant for Dean Ottinger.

Jingru Feng (China) received a Master of International Law and is a candidate for a Ph.D. doctor of law at the University of International Business and Economics in Beijing, for which she is studying at Pace Law School. She was a Research Assistant for Dean Ottinger.

Douglas S. de Figueiredo (Brazil) is an LL.M. graduate of Pace Law School, received his J.D. from the Instituto Metodista Bennett in Rio de Janeiro and now is a legal practitioner in Rio. He was a Research Assistant for Dean Ottinger.

Shakeel Kazmi (Pakistan/U.S.) is a doctoral (S.J.D.) graduate of Pace Law School and presently a Fellow of the Pace Law Center for Environmental Legal Studies. He has been a Visiting Professor at Ocean University of China and Wuhan University Law School, and an Adjunct Professor at NYU Poly. He was a Research Assistant with Dean Ottinger.

Alvin K. Leong (U.S./China) is a candidate for an LL.M. degree in Environmental Law from Pace University School of Law and is a Research Assistant to Dean Ottinger. He is a J.D. graduate of New York University School of Law.

Zheyuan Liu is an LL.M. graduate of Pace Law School and received a Bachelor of Laws degree (J.D.) from the Central Institute for Judicial Police in Hebei, China. He was a Research Assistant for Dean Ottinger.

Christopher J. Riti (U.S.) received his Juris Doctor in Environmental Law (*magna cum laude*) and Masters of Law in Climate Change and Sustainable Development from Pace University School of Law. Now a regulatory consultant, he also served as Graduate Research Fellow for the Pace Center for Environmental Legal Studies and was a Research Assistant for Dean Ottinger.

Alexis Thuau (France) received a Master of Law degree from Pace Law School; a Master in Public Law at the Université Paris, Panthéon-Sorbonne; and a Master in European Affairs at the Institut d'Études Politique, Aix en Provence, France. He was a Research Assistant for Dean Ottinger.

Long Xue (China) is an LL.M. graduate of Pace Law School, received a Bachelor of Laws degree (J.D.) from the Central Institute for Judicial Police in Hebei, China and was a Research Assistant to Dean Ottinger.

Chen Yitong (China) is a Ph.D. doctoral candidate in Environmental Law at the Ocean University of China where he also received his Bachelor of Laws degree and a Master of Philosophy degree. He was a Visiting Scholar at Pace Law School and a Research Assistant to Dean Ottinger.

Preface

This book is a compilation of studies analysing renewable energy initiatives in developing and emerging economy countries, setting forth the organization of such initiatives, the factors that contributed to their success or failure, the problems encountered and the means by which such problems were or are being addressed.

The book is designed to be of assistance to countries seeking to adopt renewable energy programs, to assist the organizers of such projects and to enable them to avoid having to re-invent the solutions to the problems of countries that have adopted them. It also should be helpful in determining the needs required to establish successful projects and the pitfalls to avoid, including assessment of the kinds of renewable energy projects that might be most suitable for them.

I have been engaged in promoting renewable energy as a clean source of energy to replace polluting fossil fuel energy resources from undependable sources for more than 55 years in many capacities: as a Peace Corps Director of Programs in Latin America from 1961-1964; an elected Member of the U.S. Congress from 1965–1970; candidate for U.S. Senate from New York in 1970; re-elected Member of Congress serving 1975–1985 (chairing the Energy, Conservation and Power Subcommittee, Energy & Commerce Committee); Professor of Environmental Law at Pace University Law School 1985–1994; Pace Law School Dean 1994-1999; Pace Law School Dean Emeritus 2000 to date; member of the Board of Directors and founder/former President of the Environmental and Energy Study Institute in Washington, DC that seeks to educate the Congress and public on environmental and energy issues, focusing in recent years on affordable clean energy and climate change remediation in developing countries; and Chair of the Environmental and Energy Specialty group of the International Union for the Conservation of Nature (IUCN).

During this time, renewable energy has gone from being a relatively small, mostly demonstration part of the energy arena to being the fastest growing energy resource. Its importance has accelerated as more efficient, reliable and less environmentally problematic technologies have been developed, off-shore wind has become feasible, thin-shelled solar photovoltaics have brought prices down sharply and the price of oil has escalated with future supplies in doubt and the existence of increasing unreliability and unrest in many of the oil-producing countries.

The United States, which developed many of the renewable energy technologies, has fallen behind in utilization of these technologies, forfeiting the leadership in all the renewable energy media to China (though coal still is its dominant energy resource), and falling behind Denmark and Germany in wind utilization. The U.S. and China, however, have the dubious distinction of not having ratified or made a commitment to reduce greenhouse gas emissions under the Kyoto Protocol to the United Nations Framework Convention on Climate Change, while being the largest global emitters of greenhouse gasses.

One of the most interesting phenomena is the commencement of renewable energy and activities in some of the major oil-producing countries. Thus Abu Dhabi successfully bid to host the first international renewable energy agency devoted to advancement of renewable energy technologies, the International Renewable Energy Agency (IRENA), and established a renewable energy research and development facility with the U.S. Massachusetts Institute of Technology to work on technology advancements in its city of Masdar. Interestingly, Masdar was established as a pollution-free city powered entirely by solar photovoltaic panels and the Masdar Institute campus has very innovative outdoor air conditioning using an ancient tower circulating cold water obtained from geothermic wells and blowing cold air into the campus, thus reducing the outdoor temperature from an average of over 100 degrees Fahrenheit to just 85 degrees. Only all-electric cars are permitted in the city. And Abu Dhabi aspires to emulate the Masdar experience for the whole country. Similarly, the Arab Emirate of Qatar just hosted the most recent United Nations climate change negotiations in its city of Doha.

This study was performed with me in conjunction with the indispensable fine work of Pace Law School students and graduate Research Assistants who did the initial principal research and drafting of most of the chapters and parts of the book. Their names and academic backgrounds are detailed in the list of contributors above. At the end of each chapter and part of the book, I have tried to summarize the most important factor in the success or failure of each case study examined as 'lessons learned.' Most of the Research Assistants are natives of the countries on which they researched and wrote. Special thanks to my wife, June Ottinger, who was always encouraging and allowed me to spend most evenings for the last month working on this manuscript. Thanks, too, to Pace Law School graduate Patricia Yak, now a practitioner, who did an initial survey of renewable energy initiatives in China.

Great credit is attributable to the late Honorable Hermann Scheer,

Member of the German Bundestag from 1980 until his death in 2010. He also was President of Eurosolar (The European Association for Renewable Energy) and General Chairman of the World Council for Renewable Energy. He worked tirelessly promoting renewable energy and was instrumental in the establishment of the International Renewable Energy Agency (IRENA) in Abu Dhabi, where I was privileged to attend the initial International Assembly. He organized and conducted a 'Renewables 2004' international conference with attendance of 154 countries and more than 3,000 participants, myself included, followed by an International Conference on Renewable Energy that formally called for the formation of IRENA. His books, *The Energy Imperative: 100 Percent Renewable Now* and *Energy Autonomy: The Economic, Social and Technological Case for Renewable Energy*, laid the foundation of modern renewable energy law, making the case for the German 'feed-in-tariff' adopted by Germany and emulated around the world. This book is dedicated to his contributions, persistent efforts and vision.

Introduction

There are many valuable lessons for all countries and companies aspiring to initiate renewable energy projects to be learned from the case studies analysed in this book. I often assured my student Research Assistants that I learned far more from them than they did from me in the course of the research they performed.

As indicated previously, I have summarized the principal lessons of the factors of success or failure of each case study, the problems encountered and the measures taken to try to overcome them. There are some general conclusions that apply to all the case studies, however, that I mention here.

Surely the most important ingredient of successful projects is knowledgeable and dedicated leadership. There are many vitally important tasks to be performed for a successful project, and competent leadership is essential.

The project leadership has first to ascertain in conjunction with local citizens and officials if the project is needed, affordable, legal and acceptable; what needs are sought to be satisfied and does the project fulfill those needs most advantageously; what mapping will be required to know what sites are most suitable from the standpoint of efficiency, environmental impact, delivery, maintenance and acceptance; what technologies are most appropriate and acceptable for the locales considered; what equipment will be required and how will it be obtained, delivered and maintained; what expertise and labor will be needed, from where it will be obtained, what training will be required, what will be the standards for such training and what redress will there be for training failures; how community participants will be informed about all aspects of the project, have input into its need and design and how will their approval be determined; how will arrangements be made for the construction, or import of the relevant technology components and maintenance and repair of them; arrangements for negotiation of contractual terms and conditions particularly where outsiders are involved, making arrangements to assure that the work is done properly and complies with all construction, safety, labor and environmental laws and regulations; provides that the companies contracted take responsibility for satisfactory performance of their work and for compensation for any damages or injuries resulting from negligent failures of performance; for training local workers, contracting for and training all personnel and government officials responsible for all aspects of the work, and for making arrangements for approvals by the appropriate government officials; provisions to assure that the host country shares adequately in the project revenues; how jurisdictional disputes among government departments that may have overlapping jurisdictions will be resolved; and how oversight and evaluation will be managed.

There are so many tasks to be performed properly in organizing and executing a renewable energy project; errors in any of them that could lead to failures, even could be life-threatening. When projects are contemplated in low income developing countries, it is important to keep in mind that they cannot afford failures and that one failed project would likely discourage other similarly situated countries from undertaking projects. So great care in selecting, designing, obtaining participant and governmental approvals, selecting and training qualified personnel and contractors, assuring adequate protections and maintenance provisions in contracts, and executing and overseeing implementation and operation of projects, while avoiding even any hint of corruption; these factors all are critically important. If a proposed project is expected to fail any of these tests, it should not be undertaken.

The book analyses renewable energy initiatives in China, including parts on biogas, solar thermal utilization, photovoltaics and off-shore wind, with a separate chapter on China's nuclear initiatives even though in itself it is not a renewable energy source, but it impacts on the ability of China to invest in renewable energy, and chapters on renewable energy in the Philippines, Morocco, India, Brazil, Indonesia and Pakistan. These contributions were all initially researched and drafted by Pace Law School student Research Assistants. I edited and contributed to each and wrote a separate chapter on hydroelectric dams and the Three Gorges Dam project in China.

1. Case studies of renewable energy in China

Richard L. Ottinger with Chen Yitong, Long Xue and Zheyuan Liu

PART I

Introduction

Richard L. Ottinger

China is in many respects unique. The largest country in the world and a top energy producer and consumer, it has remarkably grown from having no renewable energy projects a decade ago to being the world leader in most renewable technologies today. It also is the world's top greenhouse gas emitter, and investment in renewable energy is one of its best options for reducing emissions. Its renewable energy projects thus have been placed first and at greatest length in this book. With the largest number, size and variety of renewable energy projects, China presents good examples both of very successful and potentially unsuccessful projects, including those with high risks of failure.

Resolving China's energy and related environmental problems also presents a tremendous quandary, however. China has been experiencing phenomenal economic growth. Even though its growth has slowed from more than 10 percent to around 7 percent a year, it takes an enormous amount of energy to achieve and maintain such growth rates, and China's only significant domestic fuel resource is coal, the most polluting fuel and that which produces the greatest greenhouse gas emissions. China also must feed, clothe, house, employ and otherwise maintain a staggering population of 1.35 billion people. Before it opened up its economy to markets, China used to be desperately poor and its people suffered frequent devastating floods and famines.

So in weighing whether China is right in pursuing highly risky nuclear plants or large dam projects, China must consider in each case not only the risks involved in pursuing alternative energy supplies, but also the environmental, health and survival consequences of not pursuing them. With respect to renewable energy alternatives, a major question is whether enough clean energy can be produced to avoid or greatly diminish China's dependence on coal (and petroleum to fuel its growing fleet of cars and trucks). There really is no acceptable choice for resolving these problems by reducing China's economic growth with a resulting return to human tragedy. It should be noted also that China has engaged in a very aggressive program of reducing energy demand through energy efficiency measures, not covered in this book. Its alternative energy options therefore are very limited.

China's remarkable renewable energy achievements are attributable primarily to its great wealth, its acuity in finding outside funding, and its authoritarian government that permits the government to make major policy decisions and large public project investments without any significant restrictions (though recognizing that this authoritarianism results in serious employment, environmental and human rights problems).

While China is now the largest emitter of greenhouse gasses in the world, the motivation for its huge renewable energy development was not primarily related to climate change, but to the air and water pollution that are threatening the health and wellbeing of the Chinese population and increasingly are giving rise to popular anti-pollution protests.

China has many of the cities suffering the worst air pollution in the world. As the author sees it, the prime motivation of the Chinese government for instituting clean energy programs is the concern among the Chinese leadership of public uprisings against the highest world incidence of lung disease resulting from its reliance principally on coal for fueling its industrial, electrical, and domestic heating, cooling and cooking activities. Indeed, major governments historically have often been overturned by public protest against policies affecting the health and safety of the populace and the inequities of wealth distribution that left large portions of the population bereft and endangered.

As the Chinese population has learned through television and the internet that the health of its citizens is being threatened by coal burning-related pollution, public protest is becoming evident. It is primarily for this reason that China has put a high priority on the promotion of energy efficiency and production of renewable energy.

China's amazing renewable energy progress also is largely attributable to its great emphasis on education, particularly in the engineering, science, environmental law and technology fields. The author does a lot of supervision of Ph.D. (S.J.D.) and LL.M. Chinese students, and finds their preparation to be equal or superior to that of U.S. students.

A problem in China that is shared in most developing countries and even in some developed countries is that new energy technologies appear to become more widely available in the big cities, often the most affluent areas, and seldom in the rural areas that tend to be poor, most in need of them and experiencing the departure of qualified workers. This is partially the result of having much larger markets in urban areas for expensive solar and wind programs, with large international companies moving operations into the cities; but it also is the result of insufficient attention by governments to their poorer rural areas, just recently being addressed by the government.

The last universal problem with renewable energy evident in China is the advent of corruption. Even when excellent laws are enacted, they frequently are not enforced because of the influence of very wealthy domestic and foreign country suppliers (including the United States) that are willing to pay large sums to see to it that government officials ignore environmental and social justice laws or that the laws themselves are written with sizeable loopholes that permit large economic interests to avoid their implementation of them. This advent of corruption is a problem highlighted as a priority for correction by the government of China itself. Where there are large exports of renewable energy projects, corruption can be countered by the adoption of international standards that will discourage purchasers from buying renewable energy projects that are substandard in quality, environmentally unsustainable, or produced or grown with abusive labor practices. Corruption is a major factor in China and around the world in impeding the utilization of renewable energy; it is all too frequently instigated and funded by the fossil fuel industries.

The China renewable energy case study assessments covered in this book (except for Chapter 1, Part VI, written by Richard Ottinger on large dams) were initially researched and drafted by Chinese graduate students studying at Pace, including the parts on biomass, wind, solar thermal and solar photovoltaic initiatives. The parts in this chapter on China's nuclear and large dam programs, not usually considered as renewable energy, have been included because of their relation to renewable energy with respect to resolving China's energy and environmental policies. China's geothermal, ocean wave and tide programs are still in the research and development phase and thus are not included.

PART II

Biogas programs in China

Richard L. Ottinger with Chen Yitong

A. INTRODUCTION OF BIOGAS PROGRAMS IN CHINA

1. Background

The International Energy Agency defines biogas as follows:

"Biogas, also known as biomethane, swamp gas, landfill gas, or digester gas, is the gaseous product of anaerobic digestion (decomposition without oxygen) of organic matter. In addition to providing electricity and heat, biogas is useful as a vehicle fuel. When processed to purity standards, biogas is called renewable natural gas and can substitute for natural gas as an alternative fuel for natural gas equipped vehicles. Biogas is usually 50% to 80% methane and 20% to 50% carbon dioxide with traces of gases such as hydrogen, carbon monoxide, and nitrogen. In contrast, natural gas is usually more than 70% methane with most of the rest being other hydrocarbons (such as propane and butane) and traces of carbon dioxide and other contaminants. Biogas is a product of decomposing organic matter, such as sewage, animal byproducts, and agricultural, industrial, and municipal solid waste.[1]"

As a renewable energy resource, biogas is more environmentally friendly than many other traditional kinds of energy, including oil and coal.

2. How Is It Initiated and Developed in China?

China's biogas programs were started in the 1960s. At first, biogas programs were only used for domestic energy, including cooking and lighting in undeveloped rural areas. Most of these biogas programs were small household biogas digesters. China's government promoted household biogas digesters positively at that time, and formulated a nationwide technical standard for constructing and operating household biogas digesters in 1984. Since then, biogas became one of the most commonly used forms of domestic energy. Other energy resources frequently used in rural areas

---

[1]   International Energy Agency, 'Key World Energy Statistics', (2011), http://www.iea.org/textbase/nppdf/free/2011/key_world_energy_stats.pdf (accessed 11 July 2012).

20 years ago were burning firewood, and coal, both of which have severe deleterious environmental, health, equity and economic consequences.[2]

With the Chinese government's and people's awakening of environmental consciousness, biogas programs have been more and more used, with the aim of saving future energy resources and reducing pollution. The stage of development of China's biogas programs is now in the transition from millions of small household biogas digesters to large commercialized biogas programs which can supply biofuels for vehicles and generate electricity. So far, from the perspective of amount, China's household biogas digesters are extensive; however, the number of large and medium-sized biogas programs is quite limited.[3]

At the end of 2010, there were 73,000 biogas industrial projects nationwide,[4] and more than 30 million household biogas digesters.[5] The quantity of annual use of biogas was 16.5 billion cubic meters, equivalent to saving 25 million tons 'standard coal equivalent',[6] reducing carbon dioxide by more than 50 million tons.[7] The biogas programs with independent intellectual property rights in China have already been used and promoted in many developing countries in Asia and Africa, including Vietnam, Pakistan, Rwanda, Tunisia, Laos, Burma, Bangladesh and Nepal.[8]

---

[2] Chinese people have a long history of using coal as domestic energy since the Han dynasty (206 BC–AD 220). 'Coal Industry in China' [Chinese], http://baike.baidu.com/view/268353.htm (accessed 11 July 2012).

[3] Liu Zhixiong, He Xiaolan, 'Comparison and Analysis on the Development of China's Biomass Energy in the Background of Low-Carbon Economy' [Chinese] (2012) (1) Ecological Economy, http://www.gxi.gov.cn/gjtg/zjsd/zj/liuzx/201209/P020120918354004874611.pdf (accessed 1 October 2012).

[4] Pan Gaoying, Guo Yuzhi, 'Biogas is Hot and Biogas Industry is Popular, Supported by Policy and Chased by Capital' [Chinese], Shanghai Securities News (Shanghai, 24 February 2012) 4 (accessed 1 September 2012).

[5] Chen Zhengcai, 'There are more than 30 million household biogas digesters in China' [Chinese], China Green Times (Beijing, 25 August 2011) 5, http://www.greentimes.com/green/news/yaowen/zhxw/content/2011-08/25/content_143986.htm (accessed 11 July 2012).

[6] China typically converts all its energy statistics into 'metric tons of standard coal equivalent' (tce), a unit that bears little relation to the heating value of coals actually in use in China. Each tce equals 29.31 GJ (low heat) equivalent to 31.52 GJ/tce (high heat). National Research Council, Chinese Academy of Sciences, Chinese Academy of Engineering, Cooperation in the Energy Futures of China and the United States (2000), The National Academies Press, http://www.nap.edu/openbook.php?record_id=9736&page=92 (accessed 11 July 2012).

[7] Pan Gaoying, Guo Yuzhi, supra note 4.

[8] Chen Zhengcai, supra note 5.

3.　Analysis of Government Involvement and Financial Incentives

During the time of the Twelfth Five-Year Guideline (2011–15),[9] China expanded its support of biogas programs, promoting them to become another rapidly developing renewable energy resource, following behind the fast pace of solar energy and wind power growth. Promoting biogas programs was listed in the new 'Development Plan for Chinese Modern Agriculture'[10] as a special project, stating that it resulted in 'accelerating the construction of domestic energy, medium size and large scale biogas programs, and improving service for programs and advancing associated technology.' It listed that 'the popularizing rate of biogas programs in rural areas should develop from 33% in 2010 to above 50% in 2015.'[11] The government's aim for its biogas program is to deepen the industrial chain and extend the household application domain. Its aim is to promote the

---

[9]　'The five-year plans of People's Republic of China are a series of social and economic development initiatives. Planning is a key characteristic of centralized, communist economies, and one plan established for the entire country normally contains detailed economic development guidelines for all its regions. In order to more accurately reflect China's transition from a Soviet-style planned economy to a socialist market economy (socialism with Chinese characteristics), the name of the 11th five-year program was changed to "guideline" instead of "plan". The proposal for the Twelfth Five-Year Guideline was released following the fifth plenum [Chinese communists usually use the word "plenum" to replace the word "meeting" or "conference", when it refers to communist's plenary meeting] of the 17th Chinese communist party central committee on 18 October 2010, and approved by the National People's Congress on 14 March 2011, with goals of addressing rising inequality and creating an environment for more sustainable growth by prioritizing more equitable wealth distribution, increased domestic consumption, and improved social infrastructure and social safety nets. This plan is representative of China's efforts to rebalance its economy, shifting emphasis from export investment towards consumption and development from urban and coastal areas toward rural and inland areas – initially by developing small cities and greenfield districts to absorb coastal migration. The plan also continues to advocate objectives set out in the Eleventh Five-Year Plan to enhance environmental protection and accelerate the process of opening and reform.' http://en.wikipedia.org/wiki/Five-year_plans_of_the_People's_Republic_of_China#Twelfth_Guideline_.282011. E2.80.932015.29 (accessed July 11, 2012).
[10]　This 'Development Plan for Chinese Modern Agriculture' was formulated by the Ministry of Agriculture of China, and was officially announced on February 13, 2012, by the State Council. The Central People's Government of the People's Republic of China, 'Notice about Issuing the "Development Plan for Chinese Modern Agriculture (2011–2015)" by State Council' [Chinese] (The State Council, 13 February 2012) http://www.gov.cn/zwgk/2012-02/13/content_2062487.htm (accessed 11 July 112012).
[11]　Ibid.

transition of biogas from household use derived from animal and human excrement, to the transformation of all municipal solid waste to biogas for industrial use, generating electricity and connecting with natural gas pipelines.[12]

According to the 'Assessment and Evaluation methods for Methane construction and application in rural area (trial version)'[13] formulated in 2011 by the Ministry of Agriculture of China, the government will institute administrative measures to assess and evaluate associated energy administrative departments from cities and counties, enabling them to assume the responsibility of constructing biogas digesters funded by the budget of the central government. This evaluation will examine the investment and expenditure for biogas digesters, the service and application conditions, and the performance of biogas under 'normal use' periods, meaning for eight months of the year in the southern area and six months a year in northern and high altitude regions. Furthermore, in terms of subsidy, the central government will give priority to areas that implement well, and will reduce investment to other areas having poor implementation.[14] As regards the provinces with poor implementation, the Ministry of Agriculture and the National Development and Reform Commission will negotiate together and finally determine the reduction of subsidies by no less than 20 percent of the original amount received from the central government.

B. CASES OF LOCAL LEVELS

1. Yunnan Province

Yunnan Province is one of the main provinces using biogas programs. There are more than 2.5 million household biogas digesters in Yunnan Province, and almost 10 million people living in rural areas benefit

---

[12] Pan Gaoying, Guo Yuzhi, 'There Will Be Invested More Than 20 Billion RMB in the Twelfth "Five Year Guideline" (2011–2015)', Farmers Daily (Beijing, 29 February 2012) 8, http://szb.farmer.com.cn/nmrb/html/2012-02/29/nw.D110000nmrb_20120229_1-08.htm?div=-1 (accessed 11 July 2012).

[13] Office of the Ministry of Agriculture of China, 'The Notice Of Publishing "The Assessment And Evaluation Methods For Methane Construction And Application In Rural Area (Trial Version)"', Science and Education Department of Ministry of Agriculture, May 17, 2011, http://www.stee.agri.gov.cn/gdxw/t20110531_844308.htm (accessed 11 July 2012).

[14] Ibid.

from this clean renewable energy. By using biogas, the total rural area in Yunnan Province saves 4 million tons 'standard coal equivalent,' and reduces carbon dioxide by 8 million tons every year.[15] The Yunnan government has invested a significant amount of funds in the household biogas programs to protect the environment and provide an alternative energy source for farmers. During the past years, the government provided a subsidy of 1,500 RMB ($241.05) for the construction of each household biogas installation.[16]

Although the Yunnan government has invested heavily in construction, however, the maintenance cost for household biogas digesters always needs a large amount of money. In Yunnan Province now, each village has a service station that typically is in charge of 300–500 households, but since many farmers may live far away from the service station, few are willing to go to the service station to purchase a new biogas stove or fix a broken appliance. This is not only a regional problem of Yunnan Province; instead, it is a common problem all over China's rural areas. Every year, many biogas digesters are broken and abandoned because of lacking sufficient maintenance. So, how does the Yunnan government deal with this challenge?

In Yunnan Province, each station is staffed with biogas technicians to teach farmers how to properly manage the biogas digesters so as to prevent injuries. But according to Ms. Li, a government energy official of Yunnan's An Ning City, most of the biogas technicians are not paid a very high salary. She is working on introducing a policy to increase the salary of biogas technicians to attract and retain skilled technicians, so as to ensure the proper operation of biogas digesters. One major problem is the need for yearly cleaning of the digester, that includes digging up the accumulated fermented material and adding a new batch of feedstock.

Mr. Zhang Mu, the Yunnan Provincial Director of Agriculture, Department of the Energy Office, frankly stated that fragmented authority is a big problem in the Chinese political system. There are four bureaus in charge of biogas: the agricultural bureau, forestry bureau, the poverty

---

[15]   Chen Zhengcai, supra note 5.

[16]   Xiao Jinqin, 'The Construction Of Biogas Programs in Linxiang District of Yunnan Province Showed The New Picture Of Rural Area', Agriculture Development of China, 25 January 2011, http://www.farmers.org.cn/Article/ShowArticle.asp?ArticleID=92115 (accessed 11 July 2012); Liao Zhanghong, 'The Construction Of Biogas Programs in Xindian Township Is In Full Wing' [Chinese], The Window for Baihetan, March 7, 2011, http://bht.yn.gov.cn/Article/2011/20110307190402.html (accessed 11 July 2012).

alleviation bureau and the women's federation. While all the bureaus are involved in some aspects of biogas, there is no information sharing or communication between the different bureaus. The fragmented authority and lack of oversight prevents biogas programs from being well managed and better developed.[17]

While there are challenges to developing household biogas digesters in rural areas of China, the outlook for large scale biogas digesters is more promising. In Yunnan, there are 13 large scale biogas digesters (300 cubic meters and above) so far.

In addition, because of the close distance to South Asia, Yunnan Province has collaborated frequently with biogas programs in other South Asian countries. On July 18, 2011, an international biogas technology-training program for developing countries was opened in Kun Ming City of Yunnan Province. Many trainees from Burma, Thailand, Nepal and other South Asian countries participated in this course for 20 days.[18]

Yunnan Province has similar climatic conditions and similar large rural areas as in some developing countries of South Asia, and both have a large agriculture and aquaculture industry. So, the technology for commercial biogas programs is suitable for extensive promotion in this region, and deepen the collaboration with other South Asian countries, such as Bengal, Thailand, Burma and Nepal.[19]

2. Zhenping County in Henan Province

Zhenping County of Henan Province has a total area of 1,500 square kilometers, with a population of 940,000 people. By February 2012, the total investment in biogas programs had already risen to 600,000 RMB ($96,420). More than 36,000 household biogas digesters and 12 larger joint household biogas programs were constructed. The Zhenping government provides financial subsidies for promoting the construction of joint household biogas programs. Every household can receive a 2,000 RMB

---

[17]   Amy Zeng, 'Challenges to Rural Biogas-Maintenance, Lack of Technicians, and Fragmented Authority', Greening China, a Blog about China's Green Movement, 13 September 2010, http://greeningchina.wordpress.com/2010/09/13/challenges-to-rural-biogas-maintenance-lack-of-technicians-and-fragmented-authority (accessed 11 July 2012).

[18]   Hu Hongjiang, Tao Hongwei, 'Low Cost and Mature Technology of biogas programs in Yunnan Province' [Chinese], People's Daily Overseas Edition, 25 July 2011, http://yn.people.com.cn/GB/210654/211437/15242054.html (accessed 11 July 2012).

[19]   Ibid.

($321.40) subsidy and an additional subsidy equivalent to the size of each household's area.[20]

For example, Zhai Yudong, a farmer in Zhenping County, built a breeding farm that can breed 200 pigs every year. He built a 150 cubic meter biogas digester, processing pig manure into biogas. It is heated by solar power. He also operated a service station for biogas programs, which had some straw stalk grinders,[21] to provide biogas for 50 households in his county. Among his total investment of 350,000 RMB ($56,245), there was 250,000 RMB ($40,175) invested by himself, and the other 100,000 RMB ($16,070) was funded by Zhenping government subsidy.[22]

The training of personnel for maintaining biogas programs and providing a satisfactory service is always a tough task when developing and extending biogas programs. The government of Zhen Nan County required all technicians to attend training courses regularly every year, and made a strict rule of examination and evaluation for them, so as to improve their service skills.[23]

3.   Other Medium-sized and Large Biogas Programs

There also are many other medium-sized and large biogas programs being constructed. A large biogas program that can produce 70,000 cubic meters of biogas daily in Hailaer, a city of the Inner Mongolia Autonomous Region, has already finished its land planning approval procedure and is waiting for construction. An extra-large straw stalk biogas program in Yan Qing County of Beijing already has been implemented. A pressurizing and filling program for biogas that can produce 8 million cubic meters yearly in Zheng Zhou, a city of He Nan Province, and a number of programs that can transform municipal solid waste into biogas are being prepared in many cities.[24]

---

[20]   Yang Xiaoshen, Guo Yan, Yang Mingguang, 'Promoting the Socialized Service Mode For Biogas Program In Zhen Ping County of He Nan Province' [Chinese], Farmers Daily, Beijing, March 14, 2012, 8, http://www.farmer.com.cn/kjpd/nyst/201203/t20120315_705250.htm (accessed 11 July 2012).

[21]   A straw stalk grinder is a type of agricultural machinery, frequently used by Chinese farmers. It could easily grind corn stalk, straw stalk, peanut coat, and other crops into powder. These resources could be materials for producing biogas, through pyrolysis gasification.

[22]   Yang Xiaoshen, supra note 20.

[23]   Ibid.

[24]   Pan Gaoying, Guo Yuzhi, supra note 12.

C. BENEFITS OF BIOGAS PROGRAMS

1. Alleviating the Stress of Energy Supply and Reducing CO2 Emissions and Air Pollution

In 2009, fossil fuel energy accounted for 44 percent of domestic energy in rural areas of China, just behind energy from straw stalk and firewood. All of these energy resources are not environmentally and ecologically friendly. Fossil fuels are highly polluting and non-renewable energy, and straw and firewood are not conducive for maintaining vegetation and forests. All of these energy fuels will pollute the air quality and increase carbon dioxide emissions when burned.

Biogas is not only a renewable form of energy, but also is a clean energy that can reduce carbon dioxide and air pollution emissions. The main ingredient of biogas is methane (CH_4) that accounts for 50–70 percent, similar to natural gas. When burning a ton of coal, 2.6 tons of carbon dioxide and 0.02 tons of sulfur dioxide are on average are emitted. However, compared to this astonishing number, the greenhouse gasses emitted by biogas are practically zero.[25]

In this sense, as a renewable clean energy, the use of biogas can help solve the problem of shortages of fossil fuel energy and also can optimize the energy consumption structure, alleviate the stress of energy resource shortages in future and benefit the environment.

---

[25] There are many methodological research projects for measuring the carbon dioxide emissions of biogas. After a detailed mathematical proof and economical model argument of different authors, a basic conclusion is that, because biogas is carbon neutral compared with natural gas, gasoline, coal and other energy resources, the emissions from biogas can be regarded as zero. Some research supports this conclusion: Adrian Eugen Cioablă, Ioana Ionel, (eds.), 'Biomass Waste As a Renewable Source of Biogas Production-Experiments', Alternative Fuel (InTech 2011) Clear-Green, http://cdn.intechopen.com/pdfs/17591/InTech-Biomass_waste_as_a_renewable_source_of_biogas_production_experiments.pdf (accessed 11 July 2012); Environmental Inc., 'Emission Reduction/Removal Project Document, Biogas Digester Project', 3 November 2003, http://www.ghgregistries.ca/files/projects/prj_7671_645.pdf (accessed 11 July 2012); Mark Janssen, 'Liquefaction of Carbon Dioxide From Biogas Upgrading', Eindhoven University of Technology, 22 January 2010, http://students.chem.tue.nl/ifp33/download/Final%20Report%20MDP2%20Public.pdf (accessed 11 July 2012).

2.   Promote the Development of Low Carbon Agriculture

An eight cubic meter biogas digester can produce 25 tons of biogas from manure each year, providing for 0.2–0.33 hectometers farm utilization. The biodegradation of biogas from manure is nearly complete, and the biogas can be used as a very efficient fertilizer that is absorbed easily. Thus by using manure to produce biogas, farmers can reduce the amount of chemical fertilizer and pesticide needed for their crop production. The cost of production will decline and the chemical pollution resulting from crops will also be mitigated.

According to a research study, replacing the traditional chemical fertilizer by using biogas can save 1 percent of the petroleum and reduce 30 percent of greenhouse gas emissions.[26]

3.   Maintain the Forest Carbon Budget

Using biogas produced from an eight cubic meter biogas digester is equivalent to saving 2.5 tons of firewood, preserving 667 square meters of forest and reducing 2 tons of carbon dioxide emissions.[27] So, by using biogas, the forest carbon budget has been maintained.

4.   Living Changes

The biogas produced from an eight cubic meters biogas digester can satisfy the need of 3–5 people in a rural area for their living energy for one year. And it can also save their fuel charges of 1,500 RMB ($241.05).

In the Yang Chuan district of Chong Qing City, there are 40,000 biogas digesters, which save 60 million RMB ($9,642,000) of energy charges every year.[28] Farmers' lifestyles have also changed tremendously by using biogas. The living environment has been improved, and the poultry house also has been much cleaner than before.

---

[26]   Zhai Hui Juan, Liu Jin Ming, Wang Guan Qing, 'The Potential Capability Of Extensively Generating Power of Biogas In Inner Mongolia, From The Perspective Of Low Carbon Economics' [Chinese], 2011, 12 New Energy Industry, http://www.ehome.gov.cn/Article/UploadFiles/201202/2012021308412710.pdf (accessed 1 September 2012).

[27]   Supra note 25.

[28]   Lei Wen Xin, Luo Jian Yu, 'Farmers in Yong Chuan, Chong Qing City Got Benefits From Reducing Energy and Emission Reduction By Biogas Programs' [Chinese], Agriculture News, August 24, 2011, http://www.farmer.com.cn/kjpd/nyst/201108/t20110824_666311.htm (accessed 11 July 2012).

D. BIOGAS PROGRAMS REGISTERED UNDER CDM

Up to now, there are two biogas programs registered under the Clean Development Mechanism (CDM) program.[29] The first one is a 20,000 cubic meter extra-large biogas CDM program of Min He Animal Husbandry Joint Stock Company in Shan Dong Province. This project can produce 10.95 million cubic meters of biogas and 250,000 tons of organic fertilizer, deal with 180,000 tons of fowl manure, generate 2,190 kilowatts of electricity, and mitigate 67,000 tons of greenhouse gasses each year. According to the statement released publicly on February 10, 2012, this CDM project received $624,900 from the World Bank yearly from April 27, 2009 to April 30, 2010.[30]

The other biogas CDM program is in En Shi City of Hubei Province, registered on February 19, 2009. This project included 33,000 biogas households. The yearly certification emission reduction (CER) of this project is 5.84 tons. Each household of this project will receive 174 RMB ($28) on average each year as an emission reduction subsidy. Of this, 60 percent will be given to the peasant households, 18 percent will be used to provide technical services and 22 percent will be used to conduct supervision and management.[31]

In addition to these two cases, other programs are on their way to becoming biogas CDM programs. The CDM 'for low income families' biogas programs in rural area of Sichuan Province was ratified by the National Development and Reform Commission in December 2011.[32]

---

[29] The CDM program under the Kyoto Protocol to the United Nations Framework Convention on Climate Change permits developed countries to receive tradable credits from investments in carbon saving programs in developing countries.

[30] 'The Announcement by Min He (Animal Husbandry Joint Stock Company in Shan Dong Province) of Receiving CDM Income From Large Biogas Program' [Chinese], Security Times, 14 February 2012, http://money.163.com/12/0214/01/7Q6HT90R00253B0H.html (accessed 11 July 2012).

[31] 'En Shi City of Hubei Province Successfully Registered A Household Biogas CDM Program By United Nations', China Biogas, 4 April 2009, http://www.sourcejuice.com/1152543/2009/03/31/Eco-homes-Enshi-household-biogas-digesters-granted-United-Nations/ (accessed 11 July 2012).

[32] Wan Yao, The First Household Biogas CDM Program in Si Chuan Province Got Ratified' [Chinese], Si Chuan Daily, Sichuan News, 13 December 2011, http://sichuan.scol.com.cn/dwzw/content/2011-12/13/content_3220212.htm?node=968 (accessed 11 July 2012).

E.   CRITICAL ISSUES AND CAUSING FACTORS

1.   The Power Generation Policy Is Not Perfect and Complete

Although the Chinese central government announced a set of policies that promote renewable energy power generation and support a certain amount of subsidies, the People's Congress has not enacted a mandatory requirement for the purchase of renewable energy power generation.

In the absence of a mandatory act, power companies refuse to purchase power generated from biogas, asserting that it has a high cost. In China, it is very difficult to connect biogas with the power grid. Only 3 percent of biogas produced from breeding farms was used for power generation, and the rest of it was totally used for domestic gas supply and self-use.[33]

A specific regulation for connecting biogas to the power grid would promote the development of large industrial biogas programs and encourage breeding farms and other medium- and small-sized biogas digesters to make more biogas for industrial use.

2.   Lacking Professional and Persistent Technicians for Maintenance and Servicing

Biogas digesters and industrial biogas programs both need careful maintenance and routine attention. The majority of people living in China's rural area are elders, women and children, because many adult men have rushed to the cities to make money instead of staying in the isolated and underdeveloped rural areas. Since there is this lack of an adequate adult male labor force, elders and women take the responsibility to maintain and manage biogas digesters. But most of them do not have much professional knowledge of maintaining biogas programs, and technicians who are familiar and experienced with operating biogas programs are limited in number.

If no one maintains and examines them routinely, many biogas digesters would show a decrease in performance after two months' use, even under normal circumstances. Many local governments in China pay much attention to constructing biogas programs, rather than making them run well. Some local governments did found supporting service stations that provide technical guidance for biogas programs, but this kind of

---

[33]   Fang Shu Rong, 'The Handicap And Strategy of Biogas Industrialization In Rural Area Of China', 2 Journal of Agricultural Mechanization Research 2010, 216-219 (accessed 11 July 2012).

service station is still too limited to cover all biogas programs in a town or county.

In addition, these service stations are more like administrative departments, or are subject to or supervised by an administrative department, like the agriculture division on a county level. These stations are not companies, and employees of these stations receive a salary from the government instead of earnings from a station's profits. Although local governments built these service stations by themselves, they did not have adequate budgets to support consistent maintenance, and they did not give enough support to them after setting them up. Technicians receive a poor salary from these service stations, and they do not pay much attention to providing a better service for the farmers' biogas programs. And the stations always shrink gradually after their being set up.

Take a certain service station in Hay Ning City, Zhe Jiang Province, for example. Each time the cleaning up and maintenance of a 25 cubic meter biogas digester will cost 250 RMB ($40), but only a 100 RMB ($16) service fee is ordinarily charged to farmers.[34] Thus, the more service provided by technicians, the more financial loss would occur for the service stations.

3. People Are Sometimes Reluctant to Use Biogas

A basic conclusion is that the higher the living standard of farmers, the less they are likely to use biogas. People are more likely to use biogas when they are in poor conditions. With their income increasing, they will turn to choose natural gas and electricity, which are more convenient for them to use. Farmers feel biogas is too much trouble to use and maintain, and they also do not want to pay for the maintainance and examination of their biogas digesters.

The second reason is the farmers' unawareness of the true costs and benefits of biogas, including their environmentally friendly characteristics. The main reason that China's central government promotes and supports biogas programs is for preserving natural resources, saving traditional energy and protecting the environment. But farmers do not care too much about protecting the environment and saving the whole world's energy. They do not realize that biogas is an advantageous kind of renewable and clean energy that is environmentally friendly. Even when they know that biogas produced from manure can reduce their costs and replace chemical fertilizers, they still will care more about short-term benefits like 'quick

---

[34] Ge Yong Jin, Yang Sheng, 'To Ensure The Sustainable Development Of Biogas Industry', 28 Journal of China Biogas 2010 (accessed 11 July 2012).

grown crops,' and 'convenient electricity and natural gas,' rather than 'delicate biogas digesters.'[35]

4. The Need for Large and Medium Scale Biogas Programs instead of Small Household Digesters

As discussed above, most biogas programs in China are just used for domestic energy, although some biogas programs are used for providing heat, and a few of them are used for generating power. If China wants to achieve industrial development of biogas programs, the government should maintain a policy favoring small household biogas digesters, but investing more in large and medium scale industrial biogas programs.

The financial funds and subsidies are excessively flowing to small household biogas digesters, while the industrial biogas programs receive too little. Without investment and policy support, China's biogas programs lack the ability to realize the benefits of industrialization. With the acceleration of urbanization in China's rural areas, the demand for large and medium scale biogas programs will continually increase.

F. RECOMMENDATIONS

1. Improve and Complete Legislation and Policy for Renewable Power

In May 2012, the Renewable Power Quota Management Measures (discussion version) were distributed to local governments, associated companies and other interested groups, seeking consultations and suggestions. If everything goes smoothly, these Renewable Power Quota Management Measures will finally be published and enacted. This will be the most important legislation for renewable energy, after the central government updated the Renewable Energy Law of the People's Republic of China in 2009.[36]

According to this discussion version, the three responsible entities for

---

[35]   Tang Yun Chuan, Zhang Wei Feng, Ma Lin, 'Estimation of Biogas Production And Effect of Biogas Construction On Energy Economy', Transactions Of The Chinese Society Of Agricultural Engineering 2010, 281-288 (accessed 17 July 2012).
[36]   Xu Pei Yu, 'Renewable Power Quota Management Measures Will Come Out, And Associated Investment Would Blow Out' [Chinese], China Stock, 4 May 2012, http://company.cnstock.com/industry/top/201205/1992042.htm (accessed 11 July 2012).

establishing biomass projects will be power enterprises, power generation companies and local governments. Power generation companies will assume the mandatory responsibility of generating power; power enterprises will assume their mandatory responsibility of purchasing renewable energy; and the local governments will implement the full use of this kind of renewable energy. If each responsible party could fulfill its task of utilizing renewable energy, it would mean that the power market would be finally organized for renewable energy.

The National Energy Administration proposed renewable energy quotas in four areas in descending order of amount of renewable energy. On this basis, it determined the quotas index of renewable power for these four areas in terms of their renewable energy resources, economic aggregate, amount of power consumption and capacity of power transmission. The quota for renewable energy in these four areas is to be differentiated from 1 percent to 15 percent of the total power generation. This discussion version also required large power companies to generate at least 11 percent of their total generation from renewable energy resources. However, until now the final draft of this legislation still is 'under discussion.' And according to reliable news from a Ministry official of the National Energy Administration,[37] it is still too early to talk about issuing it. The stakeholders have diverse opinions on many specific details, especially on the percentage of renewable power that must be assumed by power enterprises as a mandatory responsibility.

2. Provide a Comprehensive Benefit for Household Digester Users

Although the government should give more support to industrialize large scale biogas programs, the funding for household digester users should also be optimized further. The construction and maintenance of household digesters should be combined with considering the income growth of farmers and promotion of agricultural products, instead of providing domestic energy solely.

When distributing funds for biogas programs, local governments should consider the farmers' preference for different kinds of energy and their living standards.

---

[37] Wu Yuanyuan, 'Diverse Opinions On The Percentage Of Quota, The Renewable Power Quota Management Measures Has Difficulty To Come Out Before The End Of This Year' [Chinese], Everyday Economical News, 27 November 2012, http://www.nbd.com.cn/articles/2012-11-27/697129.html (accessed 27 November 2012).

3. Industrial Use and a Vital Supplement for Natural Gas

Besides use of biomass as a domestic energy resource for people in rural areas, biogas should also be utilized as a vital supplement for natural gas, so as to encourage biogas use as an industrial product. There are three methods to commercialize biogas.

The first is to connect the biogas with a natural gas line. By eliminating CO2 and H2S, methane will consist of 95–97 percent biogas. This much purer biogas could be combined with natural gas and connected directly with the natural gas networks. Besides, biogas could also be used as a supplement for natural gas when the natural gas is limited in peak usage times. In this way, biogas and natural gas could be dispatched intelligently.

The second method is purifying the biogas into bio-natural gas for vehicle use. This purified bio-natural gas could be sold in commercial gas stations, and it has a significant price advantage over other gases. For example, in An Yang City, He Nan Province, the price of this kind of bio-natural gas is $0.75/litre (4.7 RMB/l), lower than $1/litre (6.35 RMB/l) for refined oil.[38]

Thirdly, biogas could also be used for the chemical industry. The main elements of biogas are CH4 and CO2, both of which are fundamental industrially used chemicals.

In conclusion, the future business model for biogas is to make it more industrial and commercial, so as to circulate it in the market just like natural gas.

G. LESSONS LEARNED

In addition to the suggestions above for resolving China's biogas problems, the following observations, both positive and negative, may be useful to countries planning to initiate similar projects.

Biogas energy derived from animal feces in China and other countries is a rural enterprise that somewhat offsets the urban energy trend and promotes the economic development of the farming areas. The biogas is employed for household use as well as in industrial and electricity production. China and other countries give various forms of subsidies for development of biogas facilities in agricultural areas. Examples are given of

---

[38]   Tian Liang, Li Qing-Lin, 'China's Development Strategy of Modern Gas Industry: a Perspective of Low-carbon Economy', 33 Research of Agricultural Modernization 2012 (accessed 11 July 2012).

solar thermal installations in Yunan Province, for instance, where use of biogas serves almost 10 million people living in rural areas. There are large environmental advantages in the use of such biogas in terms of reduced carbon emissions, preservation of forests and economic advantages.

A problem with the Yuhan program is that the need to maintain the biogas digesters is time-consuming and there are too few servicing stations with skilled maintenance staff. This sometimes makes biofuels less profitable than using fossil fuels. Also, the communication of the advantages of biogas is inadequate. So there is not the rapid growth in the use of biofuels that was hoped for. Lastly, there are problems of connecting biogas to power grids that could be cured by legislation, as in the U.S., which requires non-discriminatory access for all renewable energy.

PART III

China's solar thermal utilization

Richard L. Ottinger with Long Xue

A. HIMIN SOLAR CORPORATION'S SUCCESSFUL EXPERIENCE

1. Introduction

Solar thermal utilization mainly includes solar water heaters, solar thermal power generation and solar cookers. Originally solar water heaters were produced on an industrial scale to achieve economic energy savings and emission reductions. With new technologies and products developed, solar water heaters are undergoing a revolution in all sectors in China.

Solar water heaters' environmental benefits are significant in emission mitigation of SO_2, NO_2 and greenhouse gasses. Solar water heaters not only save pollution and energy, but they are more economic than electric and gas water heaters. As a result, the solar water heaters' market share has improved remarkably. In 2010, China's production of solar water heaters was 49 million square meters, accounting for about 80 percent of the world output.[39] Today, the installation of solar water heaters in China is 168 million square meters, accounting for approximately 60 percent of

---

[39] Official website of China's renewable and clean energy, http://www.crein.org.cn/ (visited November 11, 2011). [Chinese]

the world's total.[40] By 2010, the nation's solar water heating constituted 1 percent of its energy consumption, according to the Xinhua News Agency.[41]

B. HIMIN'S SUCCESSFUL DEVELOPMENT MODEL

Established on June 5, 1995, Himin Solar Corporation has grown to be the leader in the solar thermal industry in China. Himin is capable of manufacturing a list of high technology products including solar water heaters, solar collectors, split solar water heating systems, and vacuum tubes. In addition, Himin also produces large scale products such as solar power generation, solar air conditioning, solar swimming pool heaters, and solar seawater desalination.[42] Himin's success enabled the company to attract the outside investments which encourage future growth. At the 4th International Solar Cities Congress in 2010, the Himin model was highlighted as a successful business model which has given inspiration to other solar companies including China Sunergy Co. Ltd.;[43] Tianwen Group;[44] Suntech Power[45] and Shanghai Solar Energy.[46]

C. HIMIN'S 'MICRO-EMISSION EARTH' STRATEGY FOR THE SOLAR VALLEY PROJECT

On September 16, 2010, Himin released its 'Micro-Emission Earth' strategy for the first time to promote global replication of its practiced and

---

[40]   The investigation report of China's solar energy water heaters, http://homea.people.com.cn/GB/41391/3326865.html (visited January 3, 2011). [Chinese]

[41]   Xinhua News Agency, http://www.xinhuanet.com/ (visited October 22, 2011).

[42]   Himin Solar Corporation, 'Collective solar water heater solutions', http://www.himin.com/gcxt.asp (visited 12 February 2012).

[43]   CSUN, 'Energy for future', http://www.chinasunergy.com/en/ (visited January 5, 2012).

[44]   Baoding Tianwen Group Co., Ltd., http://www.twbb.com/web/eindex.asp (visited 16 December 2011).

[45]   Corporation inspired by Himin's successful business model, Suntech Power Holdings Co., Ltd., http://ap.suntech-power.com/?lang=zh (visited 9 October 2012)

[46]   Corporation inspired by Himin's successful business model, Shanghai Solar Energy S&T Co., Ltd., http://www.enf.cn/pv/2377c.html (visited 7 June 2011).

improved micro-emission future city template.[47] With the full application of solar and other clean energy technologies, this strategy guided a pilot project in China's Solar Valley.

China's Solar Valley is located in Dezhou, Shandong Province. It is host to a massive exercise in social, economic and ecological engineering by using solar power. It is planned to cost $740 million, including $10 million to install solar lighting along roadways.[48] This bold effort aims to accomplish low carbon mitigation by skillfully applying a variety of solar-based low carbon, and micro-emission technologies, and energy efficiency, in the areas of transportation, building capacity and entertainment facilities.

The Solar Valley project has attracted about 100 domestic companies and spawned other factories including Motech Industries Inc.; JA Solar Co., Ltd.; Trina Solar; Jetion Holdings Ltd. and CSG holding Co., Ltd. Also involved is the Goldman Sachs Group, Inc.,[49] which, along with Beijing-based CDH Investments, has invested $100 million in Himin Solar Corporation.[50] More than 330 hectares in area, the Solar Valley project includes a wide range of solar technologies such as water heating and air conditioning. It is trying to develop itself as the country's center of solar thermal production, logistics, research, quality testing and tourism.[51] A study in 2010 showed that a 60,000 square-meter solar floor area in the Solar Valley can produce enough energy to light and heat a city of the same size as Dezhou.[52] Therefore, China's Solar Valley gained an international reputation as the 'green city template for the 21st century,' with exemplary significance for all future cities, and has obtained recognition from international experts and leaders.[53]

This project undoubtedly put Dezhou on the map. A mid-sized city in

---

[47] Himin Solar Corporation, 'Himin Initiates "Micro-Emission Earth" Strategy Globally', http://www.himin.com/english/News/ShowArticle.asp?ArticleID=120.

[48] 'China's Solar Valley in Dezhou', http://topic.e23.cn/Content/2010-06-21/s201062100379.html (visited 12 Jan 2012).

[49] 'An American Multinational Investment Banking and Securities Firm That Engages in Global Investment Banking, Securities, Investment Management, and Other Financial Services Primarily with Institutional Clients', http://www2.goldmansachs.com/ (visited December 25, 2011).

[50] Goldman Sachs, 'CDH Invests Nearly 100 million Used in China's Himin Solar', http://www.forbes.com/feeds/afx/2008/12/12/afx5813430.html (visited 19 September 2011).

[51] China Solar Valley, 'A Journey to the Future' [Chinese], http://www.chinasolarvalley.com/Index.html (visited 12 November 2011).

[52] China Solar Valley, 'Introduction' [Chinese], http://www.chinasolarvalley.com/ (visited January 2, 2012).

[53] 'China's Solar King', http://www.greenpurchasingasia.com/content/china%E2%80%99s-solar-king-look-basics-first (visited December 15, 2011).

northwestern Shandong Province, Dezhou was a gritty city of 600,000 with surrounding suburban sprawl housing about 5 million people.[54] The city used to be known mainly as a producer of braised chicken. Today, it touts itself as 'China Solar City', a hub for clean-tech manufacturing. In an internal report on Dezhou's economic prospects, Mayor Wu Cuiyun said the city must 'use all its strength to support Himin solar energy company.'[55] She pledged 'comprehensive preference in policy, land, capital and other areas to make it a world-class enterprise group.'[56]

As part of the project, tens of thousands of farmers have been moved to alternative sites. Their land is being converted into what Himin and Dezhou's planners hope will be China's clean-technology answer to California's Silicon Valley.

D.   HIMIN'S COMPLETE INDUSTRIAL CHAIN

China has acquired independent intellectual property rights to Himin's major solar water heaters.[57] It has formed a complete industrial chain, as well as a service chain, for raw materials for solar collectors and water heater product-marketing. According to the report, around 800,000 people in Dezhou are employed in the solar industry, or one in three people of working age.[58] As a powerful and creative corporation in China, Himin provides a wide range of solar products for both industrial and domestic use.

Himin has received ISO9000, CE, TUV and Solar Keymark certificates for its solar water heater and split solar water heating systems.[59] China

---

[54]   'Introduction of Dezhou', Wikipedia, http://en.wikipedia.org/wiki/Dezhou (visited 28 March 2012).

[55]   Andrew Higgins, 'With the Solar Valley Project, China Embarks on a Bold Green Technology Mission,' http://www.washingtonpost.com/wpdyn/content/article/2010/05/16/AR2010051603482.html?hpid=topnew (visited 26 April 2012).

[56]   'Asia – China's Experimental Solar Valley, a Bold Step on Energy', The Boston Globe, http://www.boston.com/news/world/asia/articles/2010/06/06/china_takes_bold_step_on_energy_with_solar_valley/ (visited 5 June 2012).

[57]   'More Than 95% Rate of Independent Intellectual Property Rights, Brief Introduction to China Solar Valley', http://en.texpopark.com/Article/ShowArticle.asp?ArticleID=147 (visited 20 August 2011).

[58]   'China Building Ambitious "Solar Valley City" to Advance Solar Industry', http://leadenergy.org/2010/04/china-building-solar-valley/ (visited 2 January 2012).

[59]   The split solar water heating system has solar collectors (heat pipe or U-pipe) on the roof and water storage tanks inside the basement or attic; it transfers heat energy from the solar collectors into the tank via pumped circulation of the working fluid. The split system is widely used in all climates, and particularly

has adopted 17 original national level standards concerning solar water heaters, three revised in 2008 and four new standards added subsequently. Two national-level product test centers were first established: one a National Solar Water Heater Quality Supervision and Inspection Center in Beijing, and the other a National Solar Water Heater Product Quality Supervision and Inspection Center in Wuhan. And three additional solar water heater certification centers were built: (1) a China General Certification Center (CGC)[60] – Golden Sun signs;[61] (2) a China Academy of Building Research Institute (certification) – CABR signs;[62] and (3) a China Environmental United Certification Center Co., Ltd. – sign of Shihuan.[63] The first two 'National' agencies focus on the main issue of solar water heater product quality certification, while the other agency deals with the environmental impact certification of solar water heaters.[64]

---

in cold regions without risk of freezing. 'Split Solar Water Heating System,' Himin.com company website, http://www.himin.com/english/News/ShowArticle. asp?ArticleID=20 (visited 8 August 2011).

[60] Organized by the National Institute of Metrology (NIM) with the authorization of the Certification and Accreditation Administration of the People's Republic of China (CNCA), http://www.cnca.gov.cn/cnca/ (visited January 1, 2012). [Chinese]

[61] The 'China Solar Water Heater Certification System' was established under the support of the National Development and Reform Commission, the United Nations Development Program, and the United Nations Global Environment Facility. Also, the Certification System is a joint program between the Chinese Renewable Energy Industries Association and the Solar Thermal Professional Committee of China Rural Energy Industry Association. The Beijing Jianheng certification center began to implement a solar water heater certification system in 2003. Certified solar water heater products are labeled with a 'Golden Sun' certification mark, www.cgc.org.cn/ (visited October 9, 2011). [Chinese]

[62] CABR is responsible for the development and management of the major engineering construction and product standards of China. It exercises quality supervision and conducts tests on engineering construction, air conditioning equipment, solar water heaters, elevators and chemical building materials, http://www.cabr.com.cn/ (visited 11 December 2011). [Chinese]

[63] The Ltd. sign of Shihuan is displayed on the product or its packaging as a 'proof of the trademark.' It shows not only that the product is of acceptable quality, but also that it meets specific environmental requirements. Compared with other energy products, solar thermal energy is low toxicity, causes less harm, saves resources, and is more eco-friendly. Products can be certified, including: office equipment, building materials, household appliances, daily necessities, vehicles, furniture, textiles, footwear, etc., http://www.sepacec.com/ (visited 15 September 2011). [Chinese]

[64] The State Environmental Protection and Administration Authentication Center, http://www.10huan.com/news/2006-1/200613133751.html (visited 20 September 2011). [Chinese]

As a competitive corporation in China, Himin meets the demands of all the above certification centers.[65] The potential pressures on the other competitors promote the healthy development of the whole domestic market.

E. NATIONAL GOVERNMENT'S INCENTIVE POLICIES

The remarkable development of the Himin Solar Corporation might never have happened without the support and encouragement of national policies and financial incentives. The national and local governments have published several incentive policies to promote solar thermal facilities. For example, 'Solar water heater mandatory installation policies of new buildings' and 'Tax incentives policies on strengthening the solar thermal use technology research and development and promoting industrial development'[66] have been introduced in some provinces. Hainan Province is one of them. The Jiangsu Provincial Department of Construction has issued a notice to cities and counties across the province. According to that notice, any new construction built after January 1, 2007 is required to install solar water heating systems and implement building integration technology if it is a residential building or hotel under 12 stories. The solar water heating systems should coordinate with architectural design, so as to achieve the ultimate goal of synchronization. Other public buildings or residential constructions above 12 stories are also encouraged to make full use of solar water heating systems under reasonable economic conditions.[67]

In line with the requirements for energy-saving buildings, solar water heaters have a broad market domestically and internationally. Under the delegation and supervision of relevant departments, experts and scholars are drafting the standard protocols relating to solar energy and the design of architecture. A series of integrated solar water heater and architecture projects will soon be carried out simultaneously in different climate areas like Shanghai, Shandong and Yunnan.[68]

---

[65] 'A Technology Leader, Himin Solar Corporation' [Chinese], http://www. bjhimin.com/kejilingxian.aspx (visited January 5, 2012).

[66] The Central Government of People's Republic of China, http://www.gov.cn/ (visited 28 January 2012).

[67] News published by Texas neutral new energy technology Co., Ltd., http://www.sdzlxny.com/xw/xw44.html (visited 13 December 2011).

[68] The Central Government of People's Republic of China, http://www.gov. cn/ (visited 5 May 2011).

F. LESSONS LEARNED

As of 2010 China's installation of solar water heaters accounted for approximately 60 percent of the world's total and 1 percent of China's total energy consumption. The analysis of China's huge solar thermal programs fails to reveal any major problems that could be of help to other countries. They were financed with government subsidies. The only controversy over them of which the authors are aware is objections by the U.S. and other countries that subsidies granted by the government to promote all its renewable energy resources were so large as to constitute impediments to commerce violating the World Trade Organization (WTO) treaty prohibition against subsidies for exports.[69] The United Steel workers brought an action against China for its subsidies to wind energy machines that eventually was settled.[70]

PART IV

China's photovoltaic energy case study

The Shanghai 'One Hundred Thousand Solar Roofs' program

Richard L. Ottinger with Long Xue

A. INTRODUCTION

China in just a decade has gone from producing less than 1 percent of its energy from solar resources to becoming the top global producer and employer of solar thermal and solar photovoltaic (PV) energy, a truly remarkable accomplishment.[71]

Some parts of China's territories are blessed with high solar radiation resources such as Tibet, the southern part of Xinjiang, Qinghai Province, Gansu Province and the western part of Inner Mongolia. The amount of annual solar radiation in these areas is more than 1,750 kWh/m^2.[72]

---

[69] http://cleantechnica.com/2012/05/17/commerce-comes-down-hard-on-chinese-silicon-solar-pv-manufacturers/ (visited 10 April 2013).
[70] http://assets.usw.org/releases/misc/section-301.pdf/ (visited 10 April 2013).
[71] The top producers including China Sunergy Co. Ltd.; Tianwen Group; Suntech Power and Shanghai Solar Energy, see supra notes 43, 44, 45 and 46.
[72] Kan Sichao, April 2009, 'Chinese Photovoltaic Market and Industry Outlook', http://www.google.com.hk/url?sa=t&rct=j&q=annual%2Bsolar%2Bradiation%2Bin%2Bthese%2Bareas%2Bis%2Bmore%2Bthan%2B17%2B50%2BkWh%252Fm2%252E&source=web&cd=7&ved=0CGIQFjAG&url=http%3A%2F%

B. OVERVIEW OF SOLAR PV ENERGY IN CHINA

China's PV market is largely driven by the national and local govern-ments. In the beginning of the 21st century a lot of national plans were implemented including 'Power Supply Project for Un-Electrified Counties in Tibet,'[73] 'China Brightness Program,'[74] 'Brightness Program in Arli District, Tibet,' and 'Township Electrification Program.'[75]

During the period of time from China's Ninth Five-Year Plan[76] to the Eleventh Five-Year Plan,[77] several demonstration projects have been carried out, such as urban grid-connected PV electric power generation in Shandong Province and a large scale desert plant grid integration project

---

2Feneken.ieej.or.jp%2Fdata%2F3129.pdf&ei=E7--.TujyBsXh0QGygZG2BA&us g=AFQjCNGGQtu0IaPtU9jQf1iveboLcvhV8ng (visited 21 May 2011).

[73]   On July 6, 2009, the 10 megawatts photovoltaic solar power engineering project grid generation project was initiated in Xigaze, Tibet. This is the first time in the history of Shandong province that investment exceeded 100 million yuan for the solar power project of Tibet. Annually, this project has the generation capac-ity of 20.23 million kW, and it reduces 18,000 tons of carbon dioxide emissions. This project was approved by the National Development Reform Committee, http://21tyn.com/news/echo.php?id=22865 (visited 28 October 2011).

[74]   The Brightness Program includes the Township and Village Electrification Programs and is designed to bring electricity to rural areas and help alleviate poverty. China is focusing its efforts in the western provinces, including Inner Mongolia, Tibet, Qunghai, Gansu and Xinjiang, http://projects.wri.org/sd-pams-database/china/brightness-program (visited 28 December 2011).

[75]   The China Township Electrification Program (Song Dian Dao Xiang) was a scheme to provide renewable electricity to 1.3 million people (around 200,000 households) in 1,000 townships in the Chinese provinces of Gansu, Hunan, Inner Mongolia, Shaanxi, Sichuan, Yunnan, Xinjiang, Qinghai and Tibet. The program is one of the world's largest renewable energy rural electrification pro-grams. It uses a mixture of small hydro, photovoltaics and wind power. It was launched in 2001 by the State Development Planning Commission (now the National Development and Reform Commission) and was completed in 2005, http://www.google.com.hk/url?sa=t&rct=j&q=Township+Electrification+Prog ram&source=web&cd=2&ved=0CC4QFjAB&url=http%3A%2F%2Fwww.nrel. gov%2Fdocs%2Ffy04osti%2F35788.pdf&ei=gq8ET93wDMaigwfG37mxAg&us g=AFQjCNFmwOgWt7JW5b-BDM-W50JB2to-qQ (visited 18 February 2012); 'Township Electrification Program (fact sheet)', National Renewable Energy Laboratory, published April 2004, (visited May 16, 2007).

[76]   The Ninth Five-Year Plan is from 1995 to 2000 and focuses on National Economy and Social Development. It was the first medium-length plan made under a socialist market economy, and a cross-century development strategy, http://china.org.cn/95e/index.html (visited 29 May 2012).

[77]   The Eleventh Five-Year Plan is from 2006 to 2010; one of its plans is to reduce energy consumption per unit of GDP by 20%, http://www.gov.cn/english/special/115y_index.htm (visited 2 April 2012).

in Xinjiang Province. In addition, China's government has coordinated with several international private entities so as to promote the use of photovoltaic energy in rural electrification.[78]

C. THE SHANGHAI 'ONE HUNDRED THOUSAND SOLAR ROOFS' PROGRAM

Shanghai achieved an early major breakthrough in the solar energy sector causing significant impacts on domestic renewable technology innovations. Currently, there are several major solar technology programs initiated in Shanghai. Among them, the huge PV energy program, 'One Hundred Thousand Solar Roofs,' is being undertaken under the auspices of the local government. To be specific, this program was officially launched by the Shanghai Jiao Tong University Institute with the support of the World Wide Fund for Nature and the Shanghai Municipal Economic Commission,[79] at the beginning of August 2004. One Hundred Thousand Solar Roofs promotes resource savings, environmentally friendly practices and harmonious social development modes. Also, this program is subject to local conditions for solar roof planning which will promote the development of renewable energy and sustainable use of green power in Shanghai.[80]

Shanghai's solar PV power roof generation system is ideal, not only because it may help buildings achieve their maximum value, but because it brings economic benefits for local citizens as well. As a family's average monthly electricity use is 100 kilowatt hours, the electricity generated per square meter of solar roof annually may support three families' annual electricity consumption. A successful accomplishment of this program will produce 30 million kilowatts of new electricity during peak hour generation in Shanghai.[81]

---

[78] 'The Use of Photovoltaics for Rural Electrification in Northwestern China', July 1998, NREL/CP-520-23920, http://www.google.com.hk/url?sa=t&rct=j&q=China%27s+photovoltaic+energy+in+rural+electrification&source=web&cd=1&ved=0CDYQFjAA&url=http%3A%2F%2Fwww.ceibs.edu%2Fbmt%2Fimages%2F20110221%2F30210.pdf&ei=T9EET4DqJKb40gHSmsynAg&usg=AFQjCNHU4V7-n0X9RpLn8m1o6328LhzU4g (visited 22 June 2012).
[79] Shanghai Municipal Economic Commission, http://www.shanghai.gov.cn/shanghai/node17256/node17679/node17681/userobject22ai12429.html (visited 12 September 2011).
[80] 'Brief introduction of the "One Hundred Thousand Solar Roofs"' [Chinese], http://scitech.people.com.cn/GB/1057/4022108.html (visited 3 July 2010).
[81] 'The Plan of "One Hundred Thousand Solar Roofs" in Shanghai'

The plan was divided into two phases. The first phase was from 2006 to 2010, aiming to complete 10,000 solar roofs of three kilowatts each. Since each solar roof investment is calculated at 150,000 yuan, the total amount of capital required for this phase of the program is 1.5 billion yuan ($230 million).

The second phase of the program is from 2010 to 2015; it aims to complete 90,000 solar roofs, of three kilowatts each. The total amount of capital required for this phase of the program will be 8.1 billion yuan ($1.30 billion). As a whole, approximately 10 billion yuan ($1.60 billion) of investment will be involved for the One Hundred Thousand Solar Roofs program.[82]

D. NATIONAL AND LOCAL INCENTIVES RELATED TO THE 'ONE HUNDRED THOUSAND SOLAR ROOFS' PROGRAM

Renewable energy development is the key element to accomplish China's Twelfth Five-Year Plan.[83] In the first half of 2009, the government successively promulgated various policies to stimulate the domestic PV market. On March 23, 2009, the Ministry of Finance (MOF), in conjunction with the Ministry of Housing and Urban-Rural Development (MOHURD), released two new regulations: (1) the Interim Administrative Measures for Fiscal Subsidy Funds to Support the Application of Solar Photovoltaic Buildings; and (2) the Implementation Opinions on Speeding up the Application of Solar Photovoltaic Buildings to Support Demonstration Projects for Solar-powered Buildings.[84] In addition, the Renewable Energy Law of the People's Republic of China (Renewable Energy Law) was approved by the Standing Committee of the National People's Congress (NPC) and became effective on January 1, 2006.[85] Several provisions in this statute address

---

[Chinese], http://sh.eastday.com/eastday/shnews/fenleixinwen/chengjian/userobject1ai720160.html (visited 20 July 2012).

[82] Two phase construction of Shanghai's 'One Hundred Thousand Solar Roofs' plan, http://blog.eastmoney.com/tonysun111/blog_160094109.html (visited 17 March 2012). [Chinese]

[83] Deng Shasha, 'Key Targets of China's 12th Five-year Plan', March 2011, http://news.xinhuanet.com/english2010/china/2011-03/05/c_13762230.htm (visited 4 January 2011).

[84] Shanghai Economy and Information Disclosure Committee, May 31, 2011, http://www.sheitc.gov.cn/gxjscyhxdfa/511159.htm (visited December 22, 2010).

[85] The Renewable Energy Law of the People's Republic of China, 14th Session on February 28, 2005, http://www.ccchina.gov.cn/en/NewsInfo.asp?NewsId=5371 (visited 1 October 2012).

the importance of promoting the application and popularization of photovoltaic power generation and energy efficiency measures. The concept proposed in the Renewable Energy Law is similar to the feed-in-tariff mechanism pioneered in Germany. The electricity generated by solar PV panels can be sold to energy companies via a special 'feed-in' tariff.[86]

To better implement this policy, Shanghai's government set up the 'Shanghai Sunshine Fund' in order to subsidize the One Hundred Thousand Solar Roofs program in a timely manner. The Fund's project team is also in charge of solving several issues such as the associated financing arrangements, technical guidance, legal consultancy service and safety control.

E. COST-BENEFIT ANALYSIS OF SOLAR ROOF GENERATION

As previously indicated, the annual growth rate of solar PV products in all of China has increased by 20 percent annually in the past three years.[87] This growth will spur additional investment and contribute to economies of scale in solar PV manufacture. As a result, PV generation systems will be more affordable to ordinary citizens. The average life expectancy of a solar roof is about 30–50 years,[88] almost maintenance free during its operation. By taking technological progress and industrial scale factors into account, during the period 2010–15, the world's solar power costs can be reduced to 6 cents/kWh, equivalent to 0.5 yuan ($.08)/kWh or less, roughly equal to the cost of thermal power generation.[89] In conclusion, the solar roof program is both attainable and practical for citizens in Shanghai.

One of the main benefits of the use of solar roofs is it can make small scale renewable generation economically feasible for the beneficiary. Here, Shanghai's government ensures the solar roof program will bring the economic benefits for investors, producers and customers. Roughly, producers and suppliers will gain 3 percent of the average profit from the

---

[86] 'Feed-in Tariffs', Wind-works organization, http://www.wind-works.org/articles/feed_laws.html (visited 31 March 2010).

[87] 'China's Solar PV Industry Accelerating Quality Transformation', August 20, 2010, China economic net, http://en.ce.cn/Insight/201008/20/t20100820_21741172.shtml (visited 16 April 2012).

[88] Ehome, http://www.ehow.com/about_4675215_what-life-span-roof.html (visited September 26, 2011).

[89] China New Energy Chamber of Commerce, http://www.cnecc.org.cn/ (visited November 10, 2011). [Chinese]

competitive electric market.[90] Shanghai's government also provides a clear and simple basis for potential customers to calculate the returns they can expect from installation of a solar roof. Currently, the electricity price is 0.61 yuan ($.097)/kWh in Shanghai, while the solar power purchase price is 3.92 yuan ($.627)/kWh. On the basis of 3,300 kilowatts per roof, the power grid's income is 13,000 yuan per year; which means the investors can obtain a 6 percent investment return. Experts say households would normally make their money back on the cost of solar panels within about 15 years, and would continue to profit from the feed-in tariff for a further ten years. In addition, the initial users are given certain amounts of subsidies on solar roof installation. Then, the subsidy will decline annually as experience reduces the project costs.[91]

Experts from Shanghai Jiao Tong Energy Center recommended that the Shanghai Sunshine Fund should also be able to accept financial support from Shanghai's grid. It is calculated that in 2005 the city's electricity consumption had reached 96 billion kWh, and every kWh of electricity was overcharged by 0.01 yuan ($.016); therefore the city may receive nearly 960 million yuan ($153.71 million) from electric charges. This amount of money can be injected into the Shanghai Sunshine Fund so as to subsidize the One Hundred Thousand Solar Roofs program, and the budget will be balanced. In a nutshell, residents using solar power roofs receive the advantages of both social responsibility and stable returns.[92]

F.   PROGRAM CHALLENGES AND LESSONS LEARNED

1.   Lack of a Reasonable On-grid Price for Photovoltaic (PV) Generation

The Renewable Energy Law has been in effect for five years. Due to the uncertainty of pricing and pricing mechanisms, there is still no agreed upon reasonable on-grid price for PV generation.[93] According to the Renewable

---

[90]   'The Development Potential of Shanghai Solar Energy' [Chinese], http://www.biyelunwen.cn/papers/1574.html (visited 22 December 2011).

[91]   'Shanghai's Solar Roof Energy Plan' [Chinese], January 14, 2011, http://sh.people.com.cn/GB/134952/212300/212333/13736036.html (visited 10 October 2011).

[92]   'The Road of Solar Roof Utilizations in China, a Story From Mr. Zhao, President of Shanghai Solar Energy Research Institute' [Chinese], http://tech.enorth.com.cn/system/2009/07/16/004128091.shtml (visited 12 August 2012).

[93]   China PV generation's on-grid price, August 1, 2011, http://cn.reuters.com/article/chinaNews/idCNCHINA-4691520110801 (visited 7 August 2012).[Chinese]

Energy Law and its price and cost-sharing management pilot scheme, as well as the Temporary Measures of Additional Income Regulation of Renewable Energy Power, there are three main principles underlying the pricing of all renewable energy including solar energy:

a. Grid companies must purchase the full amount of renewable energy for on–grid generation;
b. Grid companies must purchase renewable energy at a reasonable price, including the reasonable cost plus a reasonable profit;
c. The costs above the regular on-grid price will be shared by all users of the national grid.[94]

More than 100 grid-connected PV generation projects have been built, but only two projects obtained an on-grid price of 4 yuan ($.64)/kWh in June 2008. Comparing PV generation prices with other renewable energy markets, wind power and biomass power generation have had clear pricing advantages in China. Besides, PV power generation's tariff approval procedures and approval process are unclear.[95]

2. Lack of Research Support

There is a variety of solar PV technology research and development units locally; however, national research funding is still limited. Sometimes the PV projects could not make substantial progress because of a serious shortage of research funding. When considering the One Hundred Thousand Solar Roofs program, instead of gaining financial research support from the central government, the initial capital was raised by the Shanghai Sunshine Fund under the support of the local government, and inadequate research funding was provided.[96]

3. Barriers to Information Exchange and Dissemination

Currently, there is no one institution that can acquire and provide PV information, either for free or not, in the PV industry. It is urgent to establish a center to provide resources, technology, financing, policy, and advisory services information for government agencies, private companies and research units related to PV industries.

---

[94] Ibid.
[95] Supra note 98.
[96] 'A Huge Construction – Shanghai "One Hundred Thousand Solar Roofs Program"' [Chinese], January 12, 2006, http://env.people.com.cn/ GB/36686/4019946.html (visited 3 February 2013).

4.   Lack of Sustainable Personnel Training Systems

China's PV industry has been making notable progress in recent years; however, the development of its corresponding training activities, training institutions, training mechanisms and authentication systems can hardly satisfy market requirements. China's PV industry will face huge obstacles in the next few decades since the rapid development of PV industries requires more professional technical and industrial expert training programs than presently are available. College degree training and public dissemination of basic science education are required to guarantee the PV industries' growth. Consequently, the Energy Research Institute under the National Development and Reform Commission suggested that authorities should: establish additional training programs and in addition accelerate the drafting of national standards for the industry; provide for PV industries to create their own standards before the nation does; and promote the creation of additional laboratory and testing facilities.[97]

5.   Overcapacity of Domestic PV Industries

Prior to 2008, China was importing about 90 percent of the raw material, polycrystalline silicon, or polysilicon, from Japan and the U.S. to make its PV panels. By mid-2008, the price of polysilicon was as high as $500 per kilogram, up from $200 per kilogram in 2007. Nevertheless, more and more domestic businesses jumped on the bandwagon to meet the international demand for PV.[98] This situation lasted until September 2008, when the global economy tanked. Statistics show that due to a sharp drop in foreign orders in 2008, China now has a photovoltaic cell production overcapacity of at least 1,000 megawatts (MW). The Chinese government is now faced with a tremendous challenge because, without sufficient overseas demand, overcapacity will threaten the profitability and long term development of China's PV industry. As a result, at the State Council working meeting on August 26, 2009, Premier Wen Jiabao pointed out that it is time for 'the redundancy in production capacity of wind energy and solar energy [to] be controlled.'[99]

---

[97]   'Solar Energy Expertise Training Program in Haining' [Chinese], http://www.360doc.com/content/10/0227/18/601991_17016511.shtml   (visited   12 November 2011).

[98]   'Overcapacity Triggers Restructuring of China's Solar PV Industry', Marketing guide, http://marketinfoguide.com/2012/01/05/overcapacity-triggers-restructuring-chinas-solar-pv-industry/ (visited 5 May 2012).

[99]   Wen Jiabao, 'Tight Monetary Policy to have a Song to Maintain Pressure',

Because of the economic crisis, China is reconsidering the development of its domestic PV industry. But one wonders why China does not use this surplus to replace its coal plants that are causing such dreadful pollution in China's cities.

Just like the One Hundred Thousand Solar Roofs program in Shanghai, the central government should implement the solar roof program in other provinces as well. This could promote the integrated application of solar-powered buildings in economically developed medium- and large-sized cities with good industrial infrastructure. In addition, it could also provide a fixed amount of subsidies to PV utilization in buildings in rural and remote regions to support the development of off-grid power generation.

PART V

China's offshore wind energy case study

The Shanghai Donghai Bridge Offshore Wind Power Demonstration Project

Richard L. Ottinger with Zheyuan Liu

A. INTRODUCTION

According to China's official report, the wind turbine market doubled every year between 2005 and 2009 in terms of total installation. It has been the world's largest annual wind market since 2009. Last year, China added 18.9 gigawatts (GW) of new wind energy capacity, meaning its total wind power installed capacity jumped to 44.7 GW, more than the United States.[100]

There are three rich wind resources in China: the North China Zone, the Coastal Wind Zone and the Tibetan Plateau. Located in the Yangtze River Delta in eastern China, Shanghai sits at the mouth of the Yangtze River in the middle portion of the Chinese coast. This location, in the Coastal Wind Zone, contains huge potential wind power. Shanghai's government began to invest in a number of environmental protection projects beginning in the 1900s.[101] Among these projects, the Shanghai Donghai

---

January 3, 2012, http://www.stockmarkettoday.cc/wen-jiabao-tight-monetary-policy-to-have-a-song-to-maintain-pressure.html (visited 27 April 2012).

[100] The Central People's Government of People's Republic of China, http://www.gov.cn/ (visited 27 December 2011).

[101] Shanghai Environmental Protection Bureau, http://www.sepb.gov.cn/fa/cms/shhj/index.htm (visited 15 December 2011).

Bridge 100 MW offshore wind power demonstration project is quite remarkable and is expected to supply electricity to power 200,000 homes for the local citizens.[102]

B. SHANGHAI DONGHAI BRIDGE SITE

Shanghai Donghai Bridge 100 MW Offshore Wind Farm Pilot Project is the first of its kind in China, built at a cost of 2.4 billion yuan (US$351 million).[103] The development and construction of the project is of great significance in improving China's offshore wind power equipment manufacturing capacity, accumulating experience in construction management of offshore wind power and promoting energy conservation and emission reduction, conservation of land resources and optimizing the energy structure in Shanghai.

The project is located in the administrative waters of Shanghai, on the east of Donghai Bridge, 8–13 kilometers from the shoreline, and with an average water depth of 10 meters. The electricity will be transmitted through a submarine cable to mainland Shanghai. 'The annual power output is expected to reach 260 million kWh, able to meet the needs yearly of over 200,000 Shanghai households. In addition, the project will save 100,000 tons of coal each year and annual carbon dioxide emissions by 200,000 tons.'[104]

The first prototype turbine of this project was hoisted and mounted on March 20, 2009. Then an initial batch of three wind turbines was put into operation on September 4, 2009. Up until August 31, 2010, 34 Sinovel SL3000 3 MW wind turbines were installed at Shanghai Donghai Bridge offshore wind farm and had successfully passed a 240-hour pre-assessment and check.[105]

---

[102] Sinovel Wind Power Technology Co., Ltd.'s website (22 September 2011). http://www.sinovel.com/en/index.aspx (visited 13 March 2012).

[103] China's Renewable Energy News, http://www.cnecc.org.cn/ (visited 2 October 2011). [Chinese]

[104] Sinovel Wind Power Technology Co., Ltd.'s website, http://www.sinovel.com/index.aspx (visited 10 September 2012).

[105] SL3000 series wind turbines are mainstream wind turbines in the world, adopting advanced technologies. This series can fully meet requirements for installations in onshore, offshore and intertidal locations. With the rated power of 3,000kW, the rotor diameter of 90/100/105/113m and the hub height of 80/90/100/110m, this series has employed multiple world-leading patented technology, and has low-voltage ride-through (LVRT) capability. (In electricity supply and generation, low voltage ride through, or fault ride through, is what an electric

The 3 MW offshore turbines were successfully mass-produced domestically, meaning China is ranked as one of the world's most advanced countries in manufacturing high-power wind turbines. Furthermore, this is the first time China has used the high-rise pile cap foundation design.[106] It is an effective solution to solve technical problems such as bearing uplift resistance and the horizontal displacement of tall wind turbines.[107]

C. SUCCESSFUL EXPERIENCE FACTORS

1. Independent Intellectual Right and Technological Merit

Sinovel Wind Power Technology Co., Ltd. (Sinovel) is a high-tech enterprise engaged in wind turbine development, design, manufacturing and marketing. Headquartered in Beijing, the company has its production base in Dalian. The company introduced 1.5 MW wind turbine manufacturing technology from abroad and developed the technology for mass

---

device, especially a wind generator, may be required to be capable of when the voltage in the grid is temporarily reduced due to a fault or load change in the grid. The voltage may be reduced in one, two or all the three phases of the AC grid. The severity of the voltage dip is defined by the voltage level during the dip and the duration of the dip.) This series is adaptable to all wind zones, environmental conditions and grid codes and requirements in the world. In November 2009 this series won the Gold Award at the China International Industry Fair 2009.

[106]   High-rise pile cap foundations are often adopted in bridges crossing rivers or straits. Pile shafts of the foundation are partly buried in the soil and partly left above the ground or the limit scour line. (Scour line is the lower elevation limit of a stream bank. It is the elevation of the ceiling of undercut bands along stream banks. On depositional banks, the scour line is the lower limit of sod-forming or perennial vegetation. On small streams, it is generally the base flow.) The riverbed scour may have a noticeable effect on pile foundations with a high-rise pile cap, and can also have a very complex influence on the seismic performance of such bridges.

[107]   Bearing capacity is the capacity of soil to support the loads applied to the ground. The bearing capacity of soil is the maximum average contact pressure between the foundation and the soil which should not produce shear failure (failure in which movement caused by shearing stresses in a soil mass is of sufficient magnitude to destroy or seriously endanger a structure) in the soil.

The uplift resistance of soils is normally associated with the design of foundations for structures such as transmission line towers where an adequate margin of safety is required against failure. Horizontal displacement is the difference between the path of the initial and final position covered by a moving object. It is calculated similarly to vertical displacement.

production.[108] At the same time it also developed models adapting to China's operating environment.

2.   Government Support

China's National Development and Reform Commission (NDRC) is a macroeconomic management agency under the Chinese State Council, which has broad administrative and planning control over the Chinese economy. Supported by the National Development and Reform Commission, the National Energy Board and relevant municipal departments, the preliminary work for the Donghai Bridge wind farm (100 MW) has been approved by the relevant government departments. According to an NDRC report, Phase II of this project will provide a pilot platform for more sophisticated wind turbines with greater unit capacity.[109]

As of August 2011 a national feed-in tariff for solar projects was enacted, and its cost is about 0.93 yuan ($ 0.15)/kWh.[110] The National Development and Reform Commission (NDRC), the country's economic planning agency, announced at the weekend four categories of onshore wind projects, which according to region will be able to apply for the tariffs. Areas with better wind resources will have lower feed-in tariffs, while those with lower outputs will be able to access more generous tariffs. The tariffs per kilowatt hour are set at 0.51 yuan ($0.082), 0.54 yuan ($0.087), 0.58 yuan ($0.094) and 0.61 yuan ($0.098).[111] Under the scheme, grid operators are forced to pay a premium for wind generation over coal power will be compensated by surcharges levied nationwide on electricity users.

On August 11, 2008, 'The Managing Measures of Wind Power Equipment Industrialization Fund' was issued by the Ministry of Finance, adopting a wind power subsidy. The approach is that a subsidy of 600

---

[108]   As China's first manufacturer that can independently develop MW-level wind turbines adaptable to all wind zones, Sinovel has succeeded in the design, development, prototype installation/commissioning/operation, GL certification and mass production of the SL1500 series of wind turbines since May 2005. Up until now, a total of over 7,000 units have been manufactured and installed, http://www.sinovel.com/en/products.aspx?ID=150 (visited 18 March 2012).

[109]   China's National Development and Reform Commission, official approval of the Shanghai Donghai Bridge Off shore Wind Power Demonstration Project, http://fgw.sh.gov.cn/main?main_colid=365&top_id=316&main_artid=8831 (visited 16 December 2011).

[110]   Yvonne Chan, 'China Sets Feed-in Tariff for Wind Power Plants', Business Green, 27 July 2009, http://www.businessgreen.com/bg/news/1801182/china-sets-feed-tariff-wind-power-plants (visited 1 January 2012).

[111]   Ibid.

yuan (approximately \$94)/kW will be paid for the first 50 sets of MW level
wind turbines manufactured. These 50 turbines must have the certification
of China General Certification (CGC) and must have been installed in
the field and put into operation.[112] That means the industrialization must
be market-accepted. The Managing Measures Fund also requires that
all turbine companies must adopt domestically made blades, gear boxes
and generators and share their subsidy with their component suppliers in
accordance with the proportion of costs.

The fact that the government accepts wind power product certification
and provides subsidies will significantly promote the domestic wind power
equipment industry.

D. MAIN PROBLEMS AND LESSONS LEARNED

1. Inadequate Mastery of the Core Technology of Wind Turbines

Currently in China, most of the domestic enterprises that are developing MW and multi-MW level wind turbines are not able to manufacture
adequately performing wind turbines of less than a megawatt (some of the
wind turbines installed at the Shanghai Donghai Bridge site are an excep-
tion). This segment of the market is still dominated by foreign enterprises.[113]

A national level public wind turbine testing facility is required, where
the overall design capacities of prototype wind turbines can be tested and
improved.

2. The Power Grid has Become a Major Bottleneck for Wind Power Development

Due to load limitations, among those wind farms that have been put into
operation, some can only feed part of their power generated into the grid

---

[112] The third-party certified by the Certification and Accreditation
Administration of the People's Republic of China, set up by the National Institute
of Metrology, http://www.cgc.org.cn/file/Orders.asp (visited 29 February 2013).

[113] 'China Wind Power Report Outlook 2011', Greenpeace, http://www.
google.com.hk/url?sa=t&rct=j&q=%E4%B8%AD%E5%9B%BD%E9%A3%8E
%E5%8A%9B%E5%8F%91%E5%B1%95%E6%8A%A5%E5%91%8A&source=
web&cd=4&ved=0CEEQFjAD&url=http%3A%2F%2Fwww.greenpeace.org%2
Fhk%2FGlobal%2Fhk%2Fpublications%2Fclimate%2Fchina-wind-power-out-
look-2011.pdf&ei=bz36Tp75K-GV0QGVw5WwAg&usg=AFQjCNF3jNATTkZ
mMGtZSKeZbQ-boVz87g (visited 4 April 2012).

even though, according to China's Renewable Energy Law, wind energy has priority access to the grid.[114]

One of the main reasons is the intermittent character of wind power; grid access is a challenge, making it difficult to meet the Renewable Energy Law requirement. One of the solutions is the establishment of a 'smart grid,' a class of technology being used to bring utility electricity delivery systems into the 21st century, using computer-based remote control and automation.[115] The design of a smart grid aims to efficiently deliver reliable, economic and sustainable electricity services. Currently smart grids are applied in several states of the U.S. and developed countries in Europe.[116]

3. The Chinese Power Grid is Monopolistic, Restricting Normal Market Competition

Grid monopolies increase the cost of operation and management and they bring additional risks.[117] An effective incentive mechanism needs to be established so as to convert the grid company's situation from a monopolistic reactive position to one that is competitive and proactive.

E. CONCLUSION

In conclusion, as the world's largest maker of wind turbines, China is making every effort to develop a green economy and be eco-friendly so as to accomplish its Twelfth Five-Year Plan. The Shanghai Donghai Bridge 100 MW offshore wind power demonstration project is an innovation which promotes renewable energy development in China. This project is a remarkable example for other developing countries with the same goals of efficient energy conservation and utilization.

---

[114]   Article 13, Chapter 4, China's Renewable Energy Law, http://www.chinanews.com/ny/news/2009/12-26/2040229.shtml (visited 27 December 2011).

[115]   U.S. Department of Energy, http://energy.gov/oe/technology-development/smart-grid (visited 3 March 2011).

[116]   Smart Grids European Technology Platform, http://www.smartgrids.eu/ (visited 19 December 2011).

[117]   Hu Xingdou, Beijing,'China's Curse: Vested Interests, Monopolies and Privileges' [Chinese], http://www.huxingdou.com.cn/urgentreforms.htm (visited 25 December 2011).

PART VI

Hydroelectric dams – Three Gorges Dam

Richard L. Ottinger

China also is relying heavily on large hydroelectric dams. China completed the world's largest hydroelectric project in terms of generating capacity, the Three Gorges Dam, on the Yangtze River in Hubei Province, fully operational in 2012. The author and colleagues from the California Energy Commission had the privilege of having a guided visit of the dam site during the last stages of its construction. China is planning to build additional hydro projects, large and small.

The Three Gorges Dam was and is very controversial.[118] It created a reservoir of 1,043.8 square kilometers, displacing 1.3 million people and many historic landmarks, farms and businesses. The dam itself is 2,335.1 meters long and 182.8 meters tall. Its 32 major generators have a capacity of 700 MW of electricity each plus two smaller generators of 50 MW, for a total production capacity at full operation of 22,500 MW, the equivalent of about 20 nuclear plants. The dam's generation displaces about 45 major coal plants, avoiding the combustion of about 100 million tons of coal per year. It also reduces what were major lethal floods from one in every 10 years to one in 100 years. It avoids about 100 million tons per year of greenhouse gasses and similarly large quantities of other dangerous pollutants from coal plants that otherwise would have been built.

The Chinese government conducted extensive public forums to inform the people who would be displaced, including markers we saw that showed the height to which the reservoir waters would rise when filled, and thus identify the displacements that would take place. Modern housing was built by the government to house all the displaced people, usually of considerably higher quality than their previous residences. Displaced businesses and farms were compensated, but usually not adequately.

On the downside of the project, the reservoir that was formed created considerable release of greenhouse gasses from deterioration of submerged vegetation and, because of submerged materials and dumping of wastes into the reservoir, it became severely polluted.[119] The wastes have accumulated at the head of the dam, impeding the flow of water for electricity production. Most seriously, the reservoir has eroded the land surrounding

---

[118]   http://www.internationalrivers.org/campaigns/three-gorges-dam, (accessed 15 November 2012).
[119]   Ibid.

the reservoir causing landslides that require the relocation of thousands of additional people. The government is paying compensation to those affected and again providing relocation housing, but this phenomenon has caused a great deal of dissatisfaction among the people affected.[120]

The Three Gorges Dam experience is a good illustration of the difficulties and conflicts involved in replacing coal-fired plants with large hydroelectric projects. It is for this reason that the World Commission on Dams was created by the World Bank and IUCN in 1998. The findings of the Commission give ample warning of the kind of problems China has experienced with the Three Gorges project and many of the other dams it is constructing.

China is now the biggest promoter of large dams, promoting not only additional dams domestically but also building some 300 large dams in 72 countries, primarily in Africa and Southeast Asia.[121]

The findings of the Commission are quite devastating as detailed in a 400-page, carefully documented report titled 'Dams and Development.' The conclusions are summarized as follows:

> Large dams have forced 40–80 million people from their homes and lands, with impacts including extreme economic hardship, community disintegration, and an increase in mental and physical health problems. Indigenous, tribal, and peasant communities have suffered disproportionately. People living downstream of dams have also suffered from water-borne diseases and the loss of natural resources upon which their livelihoods depended. Large dams cause great environmental damage, including the extinction of many fish and other aquatic species, huge losses of forest, wetlands and farmland. The benefits of large dams have largely gone to the rich while the poor have borne the costs.[122]

Smaller dam projects, particularly those utilizing run of the river electrification or the addition of hydroelectric facilities on previously existing dams are much more advantageous and are found to be acceptable renewable energy projects in the renewable energy laws of most nations.[123]

---

[120]   Sui-Lee Wee, http://www.trust.org/alertnet/news/thousands-being-moved-from-chinas-three-gorges-dam-again/ (accessed 15 November 2012).

[121]   Ibid.

[122]   http://www.internationalrivers.org/files/attached-files/wcdguide.pdf. International Rivers is an anti-dam advocacy organization, while the World Commission on Dams that researched and issued the Report is an impartial organization established by the World Bank and IUCN.

[123]   World Commission on Dams Report, ibid.

2.   Nuclear power in China: successes and challenges

Richard L. Ottinger with Jingru Feng

A.   INTRODUCTION

This analysis of China's present and planned nuclear energy development has been included separately, even though the uranium on which it relies is not inexhaustible. The authors have done so because nuclear energy plays such a large part in China's efforts to provide the energy it needs to fuel its rapid economic growth in a way designed to minimize climate change risks and other fossil fuel-derived pollution, although recognizing the risks of radioactive pollution, fish kills from once-through cooling, and environmental problems with disposal of nuclear wastes.

B.   SUCCESSFUL EXPERIENCES

In 1970, the preparation of the Qinshan Nuclear Power Plant, the first nuclear power project in China, marked the initiation of China's nuclear power path. Forty years later, China has 15[1] operating nuclear reactors producing nearly 2 percent of its power output, and there are another 27 reactors under construction, 50 more planned and more than 100 proposed. During these years, the Chinese nuclear program has witnessed rapid development and an ambitious growth plan in an effort to reduce its reliance on coal and the damage to health and the environment resulting therefrom.

1.   Support of the Central Government

The development of nuclear power plants in China mostly depends on the attitude of Chinese central government. Recently nuclear energy has been in the national spotlight.

---

[1]   Nuclear Power in China, World Nuclear Association, http://www.world-nuclear.org/, last visited October 28, 2011.

The Chinese central government sets the general future plan for national nuclear development[2]. The State Council officially issued the plan, 'Medium- and Long-term for Nuclear Power Industry from 2005 to 2020,' in which the target of nuclear power development is set clearly. By 2020, the installed nuclear capacity is expected to be over 80 GW (6%). That means nuclear power will contribute to Chinese electricity from the current 2 percent to 6 percent by 2020. A further increase to more than 200GW (16%) is expected by 2030.[3]

The Chinese government has always put nuclear safety as the priority. Following the Fukushima accident in Japan in March 2011, the Chinese government suspended its nuclear plant approval processes pending a review of lessons which might be learned from the Fukushima accident, particularly with regard to the siting of reactors, plant layouts, and control of radiation releases. Also, the Chinese government set up an independent agency specializing in the safety of nuclear power, like the U.S. Nuclear Regulatory Commission (NRC).[4]

Furthermore, the Chinese government is extremely encouraging of technical innovation and advocates domestic self-design and self-construction as well as self-operation. Premier Wen Jiabao pointed out China's nuclear development should adopt the world's advanced technologies under a unified technical line.[5]

---

[2]   "In March 2005, China's Premier Wen Jiabo called for a rapid development of the nuclear power industry. The key points of Premier Wen's announcement were: (1) Overall planning and rational layout; (2) Maximizing domestic manufacturing of nuclear power plants and equipment, with self-reliance in design and project management; (3) Encouraging international cooperation; (4) Quality and safety as a priority." United States Commercial Service, Nuclear Power Market in China, http://buyusainfo.net/docs/x_8223803.pdf, last visited November 2, 2011.

[3]   Two weeks earlier the Vice-Director of the National Development and Reform Commission (NDRC) said that China would not swerve from its goal of greater reliance on nuclear power. The former head of the National Energy Administration (NEA) said that full-scale construction of nuclear plants would resume in March 2012.

[4]   The NRC is an independent agency of the United States government and oversees reactor safety and security, reactor licensing and renewal, radioactive material safety, security and licensing and spent fuel management.

[5]   Chinese Nuclear Energy Medium and Long Term Development Plan for 2005-2020, ("4. Key Contents and Implementation of Planning"), Dynabond Powertech Service, http://www.dynabondpowertech.com/en/nuclear-power-news/topic-of-the-month/30-topic-of-the-month/2024-medium-and-long-term-development-planning-for-nuclear-power-20052020?start=4, last visited April 15, 2013

2.   Active Participation of the Local Government

The local governments are trying their best to persuade nuclear corporations to set up nuclear power plants in their own areas and provide convenient procedures for nuclear corporations, since nuclear development
could bring a huge economic profit to the local area. Take Hongyanhe,[6]
for example. The cost of all four 1080 megawatt electric (MWe) CPR-
1000 units in the first phase of construction is put at CNY 50 billion yuan
(US$7.94billion).[7] Although not all the revenues go to the local area,
there is still no doubt that the local area could acquire golden economic
effects, like employment, taxation and profits, from the future operation. Another example is Qinshan[8] Nuclear Power Plant which is located
in Haiyan County, just one of the thousands of poor counties in China
before the construction of the nuclear plant in Qinshan. However, Haiyan
has been one of the top 100 counties in China after Qinshan's operation.[9]
In 2004, the annual fiscal revenue of this small county is CNY370 million
(US$58.73 million),[10] CNY 100 million (US$15.87 million) contributed by
revenues from the nuclear industry.

---

[6]   It is the first nuclear power station receiving central government approval to
build four units at the same time and the first in northeast China.
[7]   Currencies are converted based on the recent exchange rate, At the time of
writing, 1 USD = 6.3RMB.
[8]   Qinshan is China's first indigenously designed and constructed nuclear
power plant (though with the pressure vessel supplied by Mitsubishi, Japan).
The commercial operation began in the year of 1991, owned by the China
National Nuclear Corporation, 'Qinshan Nuclear Power Station', China Culture,
http://www.chinaculture.org/gb/en_aboutchina/2003-09/24/content_26119.htm,
(last visited April 2, 2013).
[9]   The evaluation of China's top 100 counties' economies is based on three
indexes – the basic competitiveness of the county-level economy, the relative well-
to-do level and the green index. A satisfaction survey of citizens is also taken into
account. It is evaluated and released by the State Statistical Bureau, the agency
directly under the State Council, Haiyan ·China, http://english.haiyan.gov.cn/col/
col848/index.html, last visited January 20, 2012; See also Top 100 Countries in
China, Enlisting Challengle, Minhou, China Daily, http://fuzhou.chinadaily.com.
cn/e/2011-09/08/content_13651009.htm., September 8, 2011, last visited January
20, 2012.
[10]   Currencies are converted based on the recent exchange rate. At the time of
writing, 1 USD = 6.3RMB.

3. Organization of the Nuclear Industry

The organization of the Chinese nuclear industry is fragmented.[11] The nuclear industry is controlled by the Chinese government, specifically, the State Council. The China Atomic Energy Authority (CAEA) is the key agency in charge of civil nuclear programs and international cooperation; it deliberates and draws up policies and regulations on peaceful uses of nuclear energy.[12] CAEA also is the responsible institution for reviewing feasibility studies for planned nuclear power plants and radioactive waste disposal.[13] The National Nuclear Safety Administration (NNSA), under the CAEA, is the licensing and regulatory body which also maintains international agreements regarding safety. It now reports to the State Council directly.[14] The National Development and Reform Commission (NDRC) is the macroeconomic management agency directly under the State Council and is responsible for approving power projects and deciding which major nuclear power projects proceed, and when.[15] The National Energy Commission (NEC), established in March 2008, is an advisory and coordination body directly under the State Council to strengthen the government management of the energy sector.[16] It drafts a national energy development strategy complete with various programs and then monitors and implements its execution.[17] The National Energy Administration (NEA), acting as the working office of the NEC, has been established to undertake the day-to-day work of the NEC.[18] It is super-

---

[11]   Andrew C. Kadak, Nuclear Power: Made in China, Department of Nuclear Science and Engineering, Massachusetts Institute of Technology, 13 Brown J. World Aff. 77, 2006. *See also*, Government Structure and Owernship, World Nuclear Association, http://www.world-nuclear.org/info/Country-Profiles/Countries-A-F/Appendices/Nuclear-Power-in-China-Appendix-1--Government-Structure-and-Ownership/, last visited December 9, 2011.

[12]   China Atomic Energy Authority, http://www.caea.gov.cn/n360680/n360719/n360794/363641.html, December 25, 2005, last visited March 12, 2012.

[13]   *Ibid.*

[14]   http://usc.edu.cn/nec/ShowArticle.asp?ArticleID=13487, March 30, 2009, last visited March 1, 2012. Government Structure and Ownership, World Nuclear Association, http://www.world-nuclear.org/info/inf63ai_chinanuclearstructure.html, last visited February 25, 2012.

[15]   www.ndrc.gov.cn, last visited October 12, 2011. *See also* 'Government Structure and Ownership', World Nuclear Association, http://www.world-nuclear.org/info/inf63ai_chinanuclearstructure.html, http://usc.edu.cn/nec/ShowArticle.asp?ArticleID=13487, March 30, 2009, last visited March 1, 2012.

[16]   *Ibid.*

[17]   *Ibid.*

[18]   National Development and Reform Commission (NDRC) People's Republic

vised by the NDRC. The Ministry of Environment Protection (MEP) is a department directly under the State Council; it oversees health and environmental impacts and is in charge of regulations for nuclear waste disposal.[19]

Nowadays, the nuclear power industry sector is dominated by three state-owned corporations,[20] namely the China National Nuclear Corporation (CNNC), the China Guangdong Nuclear Power Group (CGNPG), and the State Nuclear Power Technology Corporation (SNPTC). The CNNC essentially controls all business aspects of the nuclear industry, including research and development, design, uranium exploration, reprocessing, and waste disposal through subsidiary companies.[21] CNNC's subsidiary, the China Nuclear Energy Industry Corporation (CNEIC), is in charge of uranium fuel trading. The CGNPG comprises some 20 companies with gross assets of CNY 133 billion (US\$21.11 billion) and net assets of CNY 41 billion (US\$6.51 billion).[22] The China Guangdong Nuclear Power Corporation (CGNPC) leads this group and is 45 percent owned by the provincial government, 45 percent by CNNC and 10 percent by CPI.[23] Another corporation, the State Nuclear Power Technology Corporation (SNPTC), has a special position. It was set up in 2004 and takes charge of technology selection for new plants acquired from overseas.[24] The SNPTC is directly under the State Council, being owned 60 percent by the State Council and with 10 percent shares owned by each

---

of China, http://en.ndrc.gov.cn/mfod/t20081218_252224.html, last visited January 20, 2011.

[19] 'Government Structure and Ownership', World Nuclear Association, http://www.world-nuclear.org/info/inf63ai_chinanuclearstructure.html, last visited January 20, 2011.

[20] Government Structure and Ownership, World Nuclear Association, http://www.world-nuclear.org/info/inf63ai_chinanuclearstructure.html#Baotau, last visited January 24, 2011.

[21] China National Nuclear Corporation, www.cnnc.com.cn., http://www.cnnc.com.cn/tabid/141/Default.aspx, last visited January 26, 2012.

[22] Government Structure and Ownership, World Nuclear Association, http://www.world-nuclear.org/info/inf63ai_chinanuclearstructure.html, last visited January 27, 2012.

[23] China Power Investment Corporation, http://eng.cpicorp.com.cn/, last visited January 27, 2012. After taking minority stakes in several early nuclear plants, CPI has now been authorized as China's third national power plant majority owner-operator, along with CNNC and CGNPG, China Nuclear Energy Industry Overview, U.S. Commercial Service, http://export.gov/china/static/732_Latest_eg_cn_026857.pdf, last visited February 2, 2012.

[24] State Nuclear Power Technology Corporation, http://www.snptc.com.cn/en/, last visited January 27, 2012.

of CNNC, CGNPC and China National Technical Import and Export Corporation.[25]

These companies are supported by institutes in China, some of which are teaching universities, others of which are design institutes. The organization of the Chinese nuclear industry relies upon the support of these institutes to provide technical input.[26] This is quite unlike that of the United States, where the prime vendors, such as Westinghouse and General Electric, oversee the design and construction of nuclear power stations and develop the license applications through to final approval by the regulator, the Nuclear Regulatory Commission (NRC). Within its government system, China is now attempting to organize its nuclear industry in a more formal way, which would permit a major expansion.[27]

4.   Adoption of High Technology

China has set the following points as key elements of its nuclear energy policy. Firstly, Pressure Water Reactors (PWRs) will be the mainstream but not sole reactor type. Secondly, nuclear fuel assemblies are fabricated and supplied indigenously. Thirdly, domestic manufacturing of plant and equipment will be maximized, with self-reliance in design and project management. Fourthly, international cooperation is nevertheless encouraged.[28]

From the beginning until now, China's nuclear power sector has been relying on technology from France, Canada, Russia and recently the U.S., with local development based largely on the French element,[29] like CNP-600 and CNP-1000.[30] So far, PWR technology mainly has been used, with three reactors designed in China (Qinshan 1, 2 and 3); six reactors purchased from France (Daya Bay, Lingao and Taishan), two reactors of Canadian design (Qinshan 4 and 5) and two Russian-designed pressurized water reactors (both VVER 1060). Qinshan 4 and 5 are Canadian

---

[25]   Government Structure and Ownership, World Nuclear Association, http://www.world-nuclear.org/info/inf63ai_chinanuclearstructure.html,   last   visited January 27, 2012.

[26]   Andrew C. Kadak, Nuclear Power: Made in China, Department of Nuclear Science and Engineering, Massachusetts Institute of Technology, 13 Brown J. World Aff. 77, 2006.

[27]   *Ibid.*

[28]   Nuclear Power in China, World Nuclear Association, http://www.world-nuclear.org/info/inf63.html, last visited November 14, 2011.

[29]   *Ibid.*

[30]   CNP-1000 is Chinese standard three-loop PWR design. CNP-600 is the two-loop PWR design. *Ibid.*

CANDU-6 pressurized heavy water reactors.[31] The latest technology acquisition has been from the U.S. (Westinghouse[32] AP 1000[33]), which is the representative of Generation III[34] technology.[35] The Westinghouse AP 1000 is the main basis of technology development in the immediate future and the main basis of China's move to Generation III technology.[36] The two plants in Sanmen and Haiyang are being equipped with the first four AP 1000 reactors.

5. Innovation and Localization

Planning for the long-term, China has also taken the lead in developing advanced nuclear technology, ensuring that the full scope of external events such as natural disasters has been taken into account, and ensuring plants can deal with multiple extreme incidents and floods in particular.[37] Meanwhile, it is reported that the main spotlights within each of the above will be plant blackout scenarios and the provision of back-up

---

[31] Eva Sternfeld, China Going Nuclear, EU-China Civil Society Forum, Hintergrund-Informationen, No. 14/2010, http://www.eu-china.net/english/Resources/Sternfeld-Eva_2010_China-Going-Nuclear.html, last visited December 1, 2011.

[32] The U.S. Export-Import Bank approved US$5 billion in loan guarantees for the Westinghouse bid. The U.S. Nuclear Regulatory Commission gave approval for Westinghouse to export equipment and engineering services as well as the initial fuel load and one replacement for four units.

[33] The Westinghouse AP 1000 is the main basis of China's move to Generation III technology, and involves a major technology transfer agreement. It is a 1250 MWe grossreactor with two coolant loops. 'Nuclear in China', World Nuclear Association, http://www.world-nuclear.org/info/inf63.html , last visited November 14, 2011.

[34] Generation II refers to a class of commercial reactors designed to be economic and reliable. Generation II systems began operation in the late 1960s and comprise the bulk of the world's 400+ commercial pressurized water reactors and boiling water reactors.General III nuclear reactors are essentially Gen II reactors with evolutionary, state-of-the-art design improvements.Stephen M. Goldberg, Robert Rosner, 'Nuclear Reactors: Generation to Generation', American Academy of Arts & Sciences, March 2011.

[35] There was a big debate on the adoption of the technology of Generation II and Generation III.Some 200 experts spent over a year evaluating Generation III designs and in September 2006 most of the 34 assigned to decide voted for the AP 1000.

[36] World Nuclear Association, http://www.world-nuclear.org/info/inf63.html, last visited December 10, 2011.

[37] Chinese Safety Checks, World Nuclear News, August 11, 2011, http://www.world-nuclear-news.org/RS_Chinese_safety_checks_1108111.html, last visited December 14, 2011.

power with appropriate redundancy.[38] Resumption of approvals for
further new plants is suspended. Moreover, a new Chinese national plan
for nuclear safety with short-, medium- and long-term actions is being
formulated, and approval for new plants will remain suspended until it is
approved.[39]

6.   Disposing of Nuclear Waste

Nuclear waste is always a hot issue in the nuclear industry. At present,
China is managing its high-, medium- and low-level nuclear waste by
on-site storage.[40] At present, medium- and low-level wastes are handled by
solidification in concrete or other materials, storage in stainless steel casks
and disposal in regional near surface disposal facilities.[41] When the regional
near surface disposal facilities are becoming full, the medium- and low-level
wastes are shipped by special truck or train to the country's storage centers
at Yumen, in Gansu Province, and Beilong, in Guangdong Province, near
the Daya Bay Nuclear Power Plant.[42] For now, high-level nuclear waste
is vitrified and stored directly at the reactor sites in the cooling pool with
boron.[43] However, in the long run, for disposal of high-level radioactive
waste, the Chinese policy is that the spent nuclear fuel should be repro-
cessed first, followed by verification and finally geological disposal.[44]

While China currently has fewer than 1,000 metric tons of high-level
spent fuel in storage, it expects to reach 2,000 metric tons by 2015. [45] Once
all the plants that are currently being ordered become operational, the
country will be producing 1,000 metric tons of high-level spent fuel per
year.[46] It is for this reason that China believes that building a national

[38]   *Ibid.*
[39]   *Ibid.*
[40]   Eva Sternfeld, China Going Nuclear, EU-China Civil Society Forum,
Hintergrund-Informationen, No. 14/2010, http://www.eu-china.net/english/
Resources/Sternfeld-Eva_2010_China-Going-Nuclear.html, last visited, last
visited December 1, 2011.
[41]   An Interview with Xue Weiming, Director of Fuel Department in CNNC,
http://news.sina.com.cn/c/2004-02-17/15392885553.shtml, February 17, 2004, last
visited December 11, 2011.
[42]   *Ibid.*
[43]   *Ibid.*
[44]   Ju Wang, High-level Radioactive Waste Disposal in China: Update 2010,
Journal of Rock Mechanics and Geotechnical Engineering, 2010, 2 (1):1–11.
[45]   *Ibid.*
[46]   Andrew C. Kadak, Nuclear Power: Made in China, Department of
Nuclear Science and Engineering, Massachusetts Institute of Technology, 13
Brown J. World Aff. 77, 2006.

Table 2.1   The 3-step long-term plan for geological disposal of High
Level Radioactive Waste (HLW) in China

Step	Period	Milestone

Step 1:
Laboratory studies
  and site selection for
  HLW repository
	2006–2020	Preliminary repository sites should be preliminary site characterizations completed. A site for an underground research laboratory (URL) is confirmed and its construction is completed. Preliminary technical capabilities in major areas.

Step 2: Underground in-site tests	2021–2040	Completion of site characterizations and confirmation of the final repository site. Completion of the most in-situ tests in the URL and establishment of technical capability for construction of the repository established. Completion of detailed repository design.
Step 3: Repository Construction	2041–2050	Completion of the repository construction around 2050. The demonstration for HLW disposal with vitrified high level waste.

repository to properly manage and dispose of the waste is extremely urgent. China has chosen deep geological disposal for the high-level waste and plans to construct a permanent disposal center to store the high-level waste.

In 2006, a three-step long-term plan for high-level radioactive waste, research and design guidelines for geological disposal of high-level radioactive waste (see the above table) was released.[47]

Since 1985, nationwide screening has been conducted to find potential repository sites. The three proposed granite sections (Juicing, Xianyangshan Xinchang and Yemaquan) are located in the Beishan region[48] in western Gansu Province, North-Western China. This area is sparsely populated and very dry, near the Gobi Desert. Several bore holes have been drilled in the area into granite formations. These exploratory studies will determine the suitability of the site for long-term disposal.

[47]   Ju Wang, High-level radioactive waste disposal in China: update 2010, Journal of Rock Mechanics and Geotechnical Engineering, 2010, 2 (1):1-11.
[48]   *Ibid.*

According to China's program, if Beishan is found to be suitable, an underground research laboratory (URL) will be built at the selected site by 2020.The URL will serve as both a methodological laboratory and a site confirmation tool. Furthermore it might be developed into an actual high-level radioactive repository.[49] Disposal is expected to start by 2050.[50]

Reprocessing[51] and recycling are mainly done in France and Japan, but in 2006, a pilot reprocessing plant with a capacity of 50 tons per year started operation in China.[52] In 2007, CNNC and French Areva signed an agreement on setting up a reprocessing plant for used fuel and mixed oxides.[53] In 2011, it was reported that Chinese scientists had mastered the technology for reprocessing fuel from nuclear power plants.[54] Although there is no detail on whether or when China is going to use this new technology on a commercial scale, China initially planned to build a commercial reprocessing plant with a capacity of 800 tons per year by 2020, but it has yet to decide on the location for the plant.[55]

However, the challenge in separated plutonium is its proliferation risk, since the separated plutonium that results from reprocessing is easily taken and transported. But it is unclear what measures China has or would adopt to protect separated plutonium from theft or export.

---

[49] *Ibid.*

[50] Wang Ju, Li Sen, Li Cheng, 'Deep Geological Disposal of High-level Radioactive Waste in China: A Three Step Strategy and Latest Progress' in Witherspoon, P.A, Bodvarsson, G.S., 'Geological Challenges in Radioactive Waste Isolation', 4th World Wide Review, Earth Science Division, Berkeley National Laboratory, USA. p.55-64 (LBNL-59808) 2006.

[51] 'Reprocessing' refers to extracting plutonium and uranium from fuel rods that have been used in a nuclear reactor. These recovered materials can be used to make new nuclear fuel, http://www.ucsusa.org/news/press_release/china-and-reprocessing-0488.html, January 13, 2011, last visited December 2, 2011.

[52] Eva Sternfeld, China Going Nuclear, EU-China Civil Society Forum, Hintergrund-Informationen, No. 14/2010, http://www.eu-china.net/english/Resources/Sternfeld-Eva_2010_China-Going-Nuclear.html, last visited, last visited December 1, 2011.

[53] *Ibid.*

[54] China Ready to Reprocess Nuclear Fuel, The Associated Press, The New York Times, January 3, 2011, http://www.nytimes.com/2011/01/04/world/asia/04china.html , last visited January 28, 2012.

[55] Yun Zhou, 'China's Spent Nuclear Fuel Management: Current Practices and Future Strategies', Center for International and Security Studies at Maryland (CISSM), Working Paper, March 2011, http://www.cissm.umd.edu/papers/display.php?id=542, last visited December 3, 2012.

7. Nuclear Export

According to Chinese officials, nuclear power is China's strategic emerging industry, and promoting civilian nuclear power exports can facilitate the upgrading and transformation of China's foreign trade structure.[56] China is now able to independently design and construct 300 MW, 600 MW and 1,000 MW nuclear power units. The China National Nuclear Corporation (CNNC), China Guangdong Nuclear Power Group (CGNPG) and the State Nuclear Power Technology Corporation (SNPTC) are developing third-generation nuclear power technology, hoping to occupy a share in the international nuclear market.

China wants to focus on exporting its civilian nuclear power plant technology to its neighboring countries, including Vietnam, Thailand, Malaysia, Belarus, Jordan and Pakistan. China has already built two 300 MW reactors for Pakistan. CNNC and the Pakistan Atomic Energy Commission are likely to enter into another agreement to conduct a joint study to finish design modifications, which would enable Pakistan to acquire two nuclear power plants, each having power generation capacity of 1,000 MG. In 2008, the Chinese government and the Jordanian government signed civilian nuclear cooperation agreements, including for the exploration of uranium, construction of the power plants and training of the work forces.

8. Personnel Training

The Chinese government has been strengthening the policy of cultivating professionals. There are six leading universities that train nuclear specialists in China and the government funds the academic institutions significantly every year. Also, the government encourages the professional staff to go abroad to study and do research with different kinds of scholarship and programs.

---

[56] This was said by Yang Qi, honorary president of the Nuclear Power Institute of China, during the Fourth Session of the 11th Chinese People's Political Consultative Conference National Committee (CPPCC) in 2011: Official: China should export nuclear power, People's Daily Online, March 2011, http://english.peopledaily.com.cn/90001/7316074.html, last visited December 5, 2012. CPPCC is a political advisory body in China. The CPPCC's main functions are to consult the-government in its policy-making and offer suggestions and criticism over its handling of various state affairs. National People's Congress Chinese People's Political Consulative Conference, http://www.china.org.cn/english/chuangye/55437.htm, last visited, December 7, 2011.

Meanwhile, the Chinese government notices the importance of the workforce safely operating and maintaining the plants. At first, China relied on French operators and engineers to assist the Chinese in operations, engineering, and reactor support, to assure that the plants' operations are in accordance with the French design and operational practices. Gradually, as the Chinese gained more operational experience, the French acted only as advisors. Now the Chinese have taken over all training responsibilities, including the training of plant operators using their on-site plant simulator at Daya Bay.[57]

The management of the China Guangdong Nuclear Power Company (CGNPC) believes that much can be learned from the U.S. plant experience to improve their overall performance. They have engaged the services of a U.S. group of experienced managers and engineers[58] to assist them in instituting U.S. processes and procedures, in order to instill the operating regime with the culture required for safe long-term operations.[59] This team spends two weeks per year at Daya Bay reviewing operating results, events, procedures, staffing levels, corrective action programs, engineering, safety culture, etc.[60] The objective is not to inspect but to mentor the Chinese staff.

9. Non-proliferation

China is a signatory to the Nuclear Non-Proliferation Treaty and the Additional Protocol under which China subjects its commercial nuclear projects to International Atomic Energy Agency (IAEA) safeguards inspections.[61] Also it signed the Comprehensive Nuclear-Test-Ban Treaty

---

[57] Eva Sternfeld, China Going Nuclear, EU-China Civil Society Forum, Hintergrund-Informationen, No. 14/2010, http://www.eu-china.net/english/ Resources/Sternfeld-Eva_2010_China-Going-Nuclear.html, last visited December 1, 2011

[58] The US requires that a senior safety executive be on duty 24-7 to handle safety problems when they arise. NRC Incident Response Plan (draft), http:// www.nrc.gov/about-nrc/regulatory/incident-response.pdf, last visited January 17, 2012.

[59] Eva Sternfeld, China Going Nuclear, EU-China Civil Society Forum, Hintergrund-Informationen, No. 14/2010, http://www.eu-china.net/english/ Resources/Sternfeld-Eva_2010_China-Going-Nuclear.html, last visited December 1, 2011

[60] *Ibid.*

[61] Stephanie Lieggi, From Proliferator to Model Citizen? China's Recent Enforcement of Nonproliferation-Related Trade Controls and its Potential, Strategic Studies Quarterly, Summer 2010, Vol.4, Issue 2, p.39, http://www.au.af. mil/au/ssq/2010/summer/lieggi.pdf, last visited March 13, 2012.

and joined the Nuclear Suppliers Group (NSG).[62] In 2003, China promulgated an export control system[63] in line with international supplier regime norms.

C. PROBLEMS AND LESSONS LEARNED

Rapid nuclear expansion such as China is planning may lead to a shortfall of qualified plant workers, safety inspectors and new challenges.

1. Shortage of Specialists and Staff

With China's rapidly expanding nuclear power industry, a potential issue has been realized – inadequate numbers of qualified professional staff. An official with the China Nuclear Society estimated the country would need 5,000 to 6,000 professionals annually in the next decade or so, versus a yearly supply now of about 2,000. If China cannot produce the necessary number of personnel, the nuclear industry will be strongly constrained by the shortage.

Although the Chinese nuclear industry has kept a pretty good safety record, with no reports of accidents of grade 2[64] or above, there have been some quality problems and accidents during plant operations, exposing gaps in management and quality control.[65] While staff can be technically trained in four to eight years, training to acquire a safety culture takes longer. This issue is also magnified by the regulatory regime, where salaries are lower than in industry, and workforce applicant numbers remain relatively low. According to the report of the State Council Research Office (SCRO)[66] in January 2011, most countries employ 30–40 regulatory staff

---

[62] China's Nuclear Fuel Cycle, World Nuclear Association, http://www.world-nuclear.org/info/inf63b_china_nuclearfuelcycle.html, last visited March 13, 2012.

[63] The system includes a 'catch-all' principle, export registration system, licensing system, end-user certification and a list control method, 'China's Non-Proliferation Policy and Measures', Information Office of the State Council of the People's Republic of China, December 2003, http://www.china.org.cn/e-white/20031202/index.htm, last visited February 26.

[64] Grade 7 is the most serious while grade 4 or lower is considered to have almost zero hazardous impact outside the plant.

[65] The comments were made by Zhang Guobao, head of the National Energy Administration.

[66] SCRO makes independent policy recommendations to the State Council on strategic matters. The Organizational Structure of the State Council, http://english.gov.cn/links.htm#1, http://www.gov.cn/gjjg/2005-12/26/content_137261.

per reactor in their fleet, but the National Nuclear Safety Administration (NNSA) has only 1,000 staff – a figure that must more than be quadrupled by 2020.

2. Regulatory Problems

The main challenges in nuclear industry regulation and supervision lie in two aspects: government organization and inadequate laws.

The organization of the Chinese nuclear industry is fragmented and overlapping. Although the China Atomic Energy Authority (CAEA) is the chief agency in charge of civil nuclear programs, there are at least ten government departments responsible for the Chinese nuclear industry, like the National Energy Agency, National Nuclear Safety Agency, the state-owned Assets Supervision and Administration Commission, the Environmental Protection Ministry, the Health Ministry and the Public Security Ministry. This overlapping regulation is always a serious potential threat to nuclear safety resulting from loose supervision and inadequate responsibility.

What is more, despite more than 20 years of deliberation, China still lacks a comprehensive nuclear industry law. Clearly, China needs to accelerate the enactment of such a law. After all, absent an overarching law codifying nuclear regulation, the dispersed nuclear governance will always raise questions about the credibility of nuclear safety in the face of expanding nuclear power and plans.

3. Investment

When talking about the investment in nuclear power plants, the Daya Bay Nuclear Power Plant's experience must be mentioned. It is the first commercial nuclear power station on the Chinese mainland and it is one of the earliest and largest joint venture projects launched under China's Open Door Policy.

However, in 1985, the building of the Daya Bay Nuclear Power Plant incited controversies and raised objections between the leaders of the Chinese government partly due to the lack of funds. Under the project, the Guangdong Nuclear Power Investment Company and the Hong Kong Nuclear Investment Company (owned by Hong Kong's China Light and Power utility) established the Guangdong Nuclear Power Joint Venture Company. Total plant construction costs for the Daya Bay project were

htm, last visited January 30, 2012.

estimated at US$4 billion. Ten percent of the US$4 billion was to be financed through equity; the other 90 percent would be financed with debt. The Hong Kong Nuclear Investment Company acquired 25 percent of the equity (equivalent to US$100 million). Guangdong Nuclear Power Investment Company subscribed US$300 million in equity. The remaining 90 percent was financed with debt provided by the Bank of China, France, the U.K. and other countries. Then the debts were supposed to be paid off with the electricity charges to customers. After coming into commercial operation, the Daya Bay Nuclear Power Plant produced around 14 billion kWh of electricity per year, of which 70 percent was sold to Hong Kong. In 2008, after 14 years of its commercial operation, Daya Bay rid itself of debt of US$5.674 billion, including all the loans and interest.[67]

Although most of the Chinese nuclear power plants follow the Daya Bay financial model, the SCRO calculated that nuclear development would require new investment of some CNY 1 trillion (US$151 billion)[68] by 2020, not counting those units being built now.[69] These projects rely mainly on debt, funds that are tight, and investment risks cannot be discounted. This cost figure could rise if supply chain issues impact schedules, with repercussions for companies borrowing to build and for the economics of the Chinese nuclear program overall. Therefore, it turns out that the Chinese government needs to find other sources of funds, not only relying on foreign investment, if China wants to satisfy the ambitious expansion of its nuclear power plant target.

4. Public Participation

When the Daya Bay nuclear power plant was built about 30 miles away from Hong Kong, it met strong public opposition in Hong Kong due to the influence of the Chernobyl disaster. More than 1 million people signed a petition to prevent the program. That was the first crisis in public relations in the Chinese nuclear industry. The Guangdong Nuclear Power Investment Company organized some Hong Kong citizens to visit the Daya Bay nuclear power plant under construction and introduced them to

---

[67] Guangdong Daya Bay NPP, http://www.cnecc.com/english/tabid/638/ctl/InfoDetail/mid/1470/InfoID/9811/language/zh-CN/Default.aspx, last visited February 12, 2012.

[68] Currencies are converted based on the recent exchange rate. At the time of writing, , 1 USD = 6.3RMB.

[69] Nuclear Power in China, World Nuclear Association, http://www.worldnuclear.org/info/inf63.html, last visited March 1, 2012.

the nuclear power plant operating experiences of the Western countries.[70]

Meanwhile, the local media made positive coverage of nuclear power. Furthermore, the Hong Kong government also visited nuclear power stations in Western countries to get an overall image of the nuclear industry.

In 2010, two minor radiation leaks in the Daya Bay plant aroused unease and debate among Hong Kong legislators, although plant management said the leaks were contained within the plant.[71]

However, the government and nuclear power plant operators have been largely untroubled by public opposition to nuclear power. People did not pay much attention before the Fukushima nuclear accident. But this is rapidly changing. Chinese citizens are more informed than in the past, and Chinese cyberspace is filled with discussion about nuclear safety.[72] People suddenly realize that China is building a lot of nuclear power plants. And they start questioning the government: do we really have a nuclear safety culture? People are not satisfied with the government's simple response of assuring that the planned nuclear power plants would be safe. Also, many officials and experts worry that the public discussions on nuclear power will mislead both the public and policymakers.[73] On the issues of public participation, the U.S. practices may provide a reference. In the U.S., the safety records and outages of nuclear power plants are required to be publicly documented. It would be helpful for China to emulate this requirement.

---

[70] CPI Jiangxi Nuclear Power Co., Ltd, http://www.jxnpc.com.cn/ReadNews.asp?NewsId=3370, February 23, 2012, last visited March 1, 2012.

[71] Kari Huus, China's Nuclear Energy Policy: Build, baby, build!, China on NBCNEWS, March 24, 2011,,http://www.msnbc.msn.com/id/42219006/ns/world_news-asia_pacific/t/chinas-nuclear-energy-policy-build-baby-build/, last visited February 19, 2012.

[72] *Ibid.*

[73] Hepeng Jia, Nuclear Debates Call for Public Participation, Advancing the Chemical Sciences, July 6, 2011, http://www.rsc.org/chemistryworld/News/2011/July/06071101.asp, last visited February 20, 2012.

APPENDIX

Table 2A.1   Operating nuclear reactors

Units	Province	Net capacity (each)	Type	Operator	Commercial operation
Daya Bay 1&2	Guangdong	944 MWe	PWR (French M310) PWR (CNP-300)	CGNPC	1994
Qinshan Phase I	Zhejiang	279 MWe		CNNC	April 1994
Qinshan Phase II, 1–4	Zhejiang	610 MWe	PWR (CNP-600)	CNNC	2002, 2004, 2010, (2012)
Qinshan Phase III, 1&2	Zhejiang	665 MWe	PWR (Candu-6)	CNNC	2002, 2003
Ling Ao Phase I, 1&2	Guangdong	935 MWe	PWR (French M310)	CGNPC	2002, 2003
Taiwan 1&2	Jiangsu	1,000 MWe	PWR (VVER-1000)	CNNC	2007, 2007
Ling Ao Phase II, 1&2	Guangdong	1,037 MWe	PWR (CPR-1000)	CGNPC	Sept 2010, Aug 2011
Total: 15		11,881 MWe			

Source: http://www.world-nuclear.org/info/inf63.html.

3. Renewable energy in the Philippines

Richard L. Ottinger with
Alvin K. Leong

A. INTRODUCTION

The Philippines has a strategic energy policy to promote the exploration, development and utilization of renewable energy resources. The Philippine Department of Energy ('PDOE') has stated:

> The harnessing and utilization of renewable energy (RE) comprises a critical component of the government's strategy to provide energy supply for the country. This is evident in the power sector where increased generation from geothermal and hydro resources has lessened the country's dependency on imported and polluting fuels. In the government's rural electrification efforts, on the other hand, renewable energy sources such as solar, micro-hydro, wind and biomass resources are seeing wide-scale use.
>
> It is the government's policy to facilitate the energy sector's transition to a sustainable system with RE as an increasingly prominent, viable and competitive fuel option. The shift from fossil fuel sources to renewable forms of energy is a key strategy in ensuring the success of this transition. Moreover, current initiatives in the pursuit of this policy are directed towards creating a market-based environment that is conducive to private sector investment and participation and encourages technology transfer and research and development. Thus, current fiscal incentives provide for a preferential bias to RE technologies and projects which are environmentally sound.[1]

In terms of installed renewable energy capacity, the Philippines is currently ranked sixth in Asia and thirty-fourth in the world.[2] In 2012,

---

[1] Renewable Energy, Energy Resources Official Website of the Philippine Department of Energy, http://www.doe.gov.ph/ER/Renenergy.htm, accessed 9 April, 2013.
[2] Renewable Facts, http://www.renewablefacts.com/country/philippines/renewables, accessed 9 April 2013.

its total renewable energy capacity represented 32.4 percent of its total installed energy capacity.[3] Its renewable energy industry is led by micro-hydroelectric power (63%) followed by geothermal power (36%).[4] The Philippines is currently the world's second largest producer of geothermal power, ranking below the United States.[5] It currently has an installed capacity of 1,902 MW of geothermal energy.[6] The Philippines has indicated that one of its goals is to become the number one geothermal power producer in the world.[7]

Currently the Philippines is considered to be the most developed renewable energy market in South East Asia.[8] The government plans to double the country's renewable energy capacity by 2030, and its specific technology goals include:

- increasing hydropower capacity by 160 percent;
- increasing geothermal capacity by 75 percent;
- building a wind power grid to provide an additional 2,345 MW of capacity;
- developing an additional 277 MW of biomass power;
- developing an additional 284 MW of solar power capacity;
- building the country's first ocean energy facility.[9]

This case study will examine the Philippine renewable energy legal and regulatory framework established by the Renewable Energy Act of 2008, with a particular focus on the implementation of the Feed-in Tariff System it has adopted.

---

[3] Ibid.

[4] 'Meeting the Energy Challenge in South East Asia,' A Paper on Renewable Energy, Ipsos Business Consulting, July 2012, http://w3.ipsos.com/businessconsulting/insights/whitepaper/docs/A_Paper_on_Renewable_Energy_in_South_East_Asia_July_2012.pdf, accessed April 9, 2013.

[5] Renewable Facts, http://www.renewablefacts.com/country/philippines/1750-philippines-to-become-largest-geothermal-energy-producer-in-the-world, accessed 9 April 2013..

[6] Ibid.

[7] Renewable Energy, Energy Resources, Energy Sector Objectives, RE Goals, Policies and Strategies, Official Website of the Philippine Department of Energy, http://www.doe.gov.ph/ER/RE%20tables%20pdf/energy%20sector%20objectives.pdf, accessed 9 April 2013.

[8] Ipsos, supra note 4.

[9] Ipsos, supra note 4.

B.   THE RENEWABLE ENERGY ACT OF 2008

The Renewable Energy Act of 2008[10] (the 'RE Law') was signed into law in December 2008.

The RE Law declared the policy of the government as three-fold:[11]

1. Accelerate the exploration of renewable energy resources to achieve energy self-reliance through the adoption of sustainable energy development strategies to reduce the country's dependence on fossil fuels.
2. Increase the utilization of renewable energy by institutionalizing the development of national and local capabilities in the use of renewable energy systems and promoting the efficient and cost-effective commercial application by providing fiscal and non-fiscal incentives.
3. Encourage the development and utilization of renewable energy resources to effectively prevent or reduce harmful emissions and thereby balance the goals of economic growth and development with the protection of health and the environment.

The RE Law's stated scope is to establish the framework for the accelerated development and advancement of renewable energy sources and the development of a strategic program to increase its utilization.[12] The RE Law designated the PDOE as the lead agency to implement the provisions of the Act,[13] and created a National Renewable Energy Board ('NREB') to, among other things, recommend specific actions to facilitate the implementation of a National Renewable Energy Program to be executed by the PDOE.[14]

In terms of on-grid renewable energy development, Chapter III of the RE Law provides five important tools:[15]

1. Renewable Portfolio Standard ('RPS') – the NREB will set the minimum percentage of generation from eligible renewable energy resources and determine to which sector RPS will be imposed on a

---

[10]   An Act Promoting the Development, Utilization and Commercialization of Renewable Energy Resources and for Other Purposes, Rep. Act No. 9513 (Dec. 16, 2008) (Phil.), Official Gazette of the Republic of the Philippines, http://www. gov.ph/2008/12/16/republic-act-no-9513/, accessed 9 April 2013.

[11]   Ibid. at § 2.

[12]   Ibid. at § 3.

[13]   Ibid. at § 5.

[14]   Ibid. at § 27.

[15]   Ibid. at §§ 6–11.

per grid basis. As of November 2012, the RPS rules have yet to be finalized.

2. Feed-in Tariff System ('FIT System') – a feed-in tariff system for electricity produced from wind, solar, ocean, run-of-river hydropower and biomass is mandated. The PDOE has promulgated rules and approved tariff rates, which will be discussed in more detail below.
3. Renewable Energy Market ('REM') – to facilitate compliance with the RPS, the PDOE is tasked with integrating the REM into the Wholesale Electricity Spot Market.
4. Green Option Energy – to provide end-users with the option to choose renewable energy resources as their sources of energy.
5. Net-metering for Renewable Energy – subject to technical considerations and upon request by distribution end-users, the distribution utilities will enter into net-metering agreements with qualified end-users.

From an investment perspective, providing for the establishment of both an RPS and a FIT System is significant. The difference between an RPS and a FIT System is essentially the difference between demand and supply: an RPS mandates the amount of electricity demand that must be met by renewable energy resources while a FIT System encourages the development of a new renewable energy supply by providing greater revenue certainty for investors.[16] FIT systems are focused on setting the right price of renewable energy to incentivize investment, while RPS policies are focused on the quantity of renewable energy, typically leaving the price up to competitive bidding.[17] There are likely to be policy interactions between the two: the absence of long-term support provided by a FIT System could result in renewable energy projects encountering difficulty in securing financing, which could lead to a shortage of supply to meet RPS demand.[18]

Chapter VII of the RE Law provides a host of commercial and financial incentives[19] for qualifying renewable energy projects and activities, including:

---

[16] See Karlynn Cory, Toby Couture and Claire Kreycik, 'Feed in Tariff Policy: Design, Implementation, and RPS Policy Interactions,' National Renewable Energy Laboratory, March 2009, http://www.nrel.gov/docs/fy09osti/45549.pdf, accessed 9 April 2013.
[17] Ibid.
[18] Ibid.
[19] Renewable Energy Act, supra note 10, at § 15.

1. Income Tax Holiday for the first seven years of commercial operations;
2. Duty-Free Importation of renewable energy machinery, equipment and materials, within the first ten years upon the issuance of a certification;
3. Special Realty Tax Rates on equipment and machinery used for renewable energy;
4. Net Operating Loss Carry-Over for RE facilities, such tax rate not exceeding 1.5 percent of the original cost less accumulated normal depreciation or net book value ('NOLCO'), where the NOLCO of the renewable energy developer during the first three years from the start of commercial operation may be carried over as a deduction from gross income for the next seven consecutive taxable years immediately following the year of such loss;
5. Reduced Corporate Tax Rate, where after seven years of income tax holiday, renewable energy developers will pay a corporate tax rate of 10 percent (compared to the normal corporate tax rate of 30%);
6. Accelerated Depreciation available under certain circumstances, where plant, machinery and equipment reasonably needed and actually used for the exploration, development and utilization of renewable energy resources may be depreciated up to twice the rate of normal depreciation;
7. Zero-Percent Value-Added Tax Rate, where the sale of fuel or power from renewable sources of energy will be subject to zero percent value-added tax; and
8. Tax Exemption of Carbon Credits, where all proceeds from the sale of carbon emission credits will be exempt from all taxes.

It can be said that the RE Law, in establishing a comprehensive framework for renewable energy development, clearly demonstrates the Philippine government's strong commitment to promoting and advancing renewable energy. The recent developments involving the implementation of the FIT System are particularly significant for renewable energy investors interested in the Philippine renewable energy markets. On July 12, 2010, the Philippine Energy Regulatory Commission ('ERC'), which is part of the PDOE, promulgated the Feed-in Tariff Rules ('FIT Rules').[20]

---

[20] Resolution Adopting the Feed-In Tariff Rules, Resolution No. 16, Series of 2010, Energy Regulatory Commission, 12 July 2010, (Phil.), Energy Regulatory Commission – Philippines, http://www.erc.gov.ph/Issuances/resolutions?pageInde x=2&displayYears=2010, accessed 9 April 2013. The FIT Rules took effect on 12 August 2010.

Two years later, on July 27, 2012, the ERC approved Feed-in Tariff Rates for hydroelectric, biomass, wind and solar energy sources.[21] Because of the importance of the FIT System to the potential growth and development of the Philippine renewable energy markets, we turn our attention to an examination of this program.

C.   THE FIT SYSTEM

Generally speaking, a feed-in tariff system works by guaranteeing fixed tariffs for electricity produced by eligible renewable energy projects for a specified timeframe. It is intended to attract investments in renewable energy projects by providing greater certainty of revenue (typically at an above-market price) for project developers, which in turn should attract project lenders into the market (as such lenders typically lend against this revenue stream). The main structural elements of the FIT Rules[22] can be summarized as follows:

1.   It applies to on-grid areas, and covers wind energy, solar energy, ocean energy, run-of-river hydroelectric power, biomass energy and renewable energy components of hybrid systems.
2.   It is a fixed tariff instead of a premium over market prices and will be set by the ERC.
3.   The ERC will adjust tariffs annually to allow pass-through of inflation and foreign exchange rate variations.
4.   The tariffs will be subject to a digression rate[23] set by the ERC.
5.   Eligible renewable energy plants (which must obtain certificates of compliance from the ERC) will be entitled to the applicable tariffs for a period of 20 years.
6.   The tariffs will cover the costs of the plant, as well as the costs of connecting the plant to the transmission or distribution network, calculated over the expected life of the plant, and provide for a

---

[21]   Resolution Approving the Feed-In Tariff Rates, Resolution No. 10, Series of 2012, Energy Regulatory Commission, 28 August 2012 (Phil.), Energy Regulatory Commission – Philippines, http://www.erc.gov.ph/Issuances/resolutions, accessed 9 April 2013.

[22]   See Annex A (Feed-In Tariff (FIT) Rules) attached to Resolution Adopting the Feed-In Tariff Rules, supra note 20.

[23]   Digression rate refers to the rate to be applied to the FITs to reduce it over time, to take into account the maturing of renewable energy technology and the resulting cost reduction. Ibid.

market-based weighted average cost of capital in determining return on invested capital.

7. Electricity consumers who are supplied with electricity through the distribution or transmission network will share in the cost of the tariffs through a uniform charge (Philippine peso ('PHP')/kWh), referred to as the feed-in tariff allowance.

8. Eligible renewable energy plants will enjoy priority connection to the transmission or distribution system.

On May 16, 2011, the NREB submitted its proposed tariff rates as follows (each in PHP/kWh):[24]

1. Biomass: 7.00
2. Ocean: 17.65
3. Hydropower: 15.00
4. Solar: 17.95
5. Wind: 10.37

The NREB also proposed a digression rate of 0.5 percent after year 2 for biomass, hydropower and wind, and 6 percent after year 1 for solar, but no digression for ocean energy.[25]

On July 27, 2012, the ERC adopted a resolution approving the following tariff rates (each in PHP/kWh):[26]

1. Biomass: 6.63
2. Hydropower: 5.90
3. Solar: 9.68
4. Wind: 8.53

The ERC deferred setting a tariff rate for ocean energy pending further study and data gathering.[27] For digression, the ERC adopted the same rates as those recommended by the NREB.[28]

As can be seen, the tariff rates adopted by the ERC are lower (and in

---

[24]   See Resolution Approving the Feed-In Tariff Rates, supra note 21. See also NREB's Petition to Initiate Rule-making for the Adoption of Feed-In-Tariff, Climate Investment Funds, https://www.climateinvestmentfunds.org/cif/sites/cli mateinvestmentfunds.org/files/NREB(Petition-FIT).pdf., accessed 9 April 9 2013

[25]   Resolution Approving the Feed-In Tariff Rates, supra note 21

[26]   Ibid.

[27]   Ibid.

[28]   Ibid.

several categories, substantially lower) than those recommended by the NREB. In its press release,[29] the ERC stated that it accepted the methodology used by the NREB in calculating its proposed tariffs, and provided the following explanations for the lower rates:

1. The ERC arrived at lower rates for wind and solar after it updated the construction costs of representative plants to reflect the 'downward market trend' of the costs of putting up these plants;
2. The ERC adopted higher capacity factors for these plants 'to ensure that only the more efficient plants' will enjoy the FIT incentive;
3. For all the applicable renewable energy technologies, the ERC revised other project costs such as those for the switchyard and transformers, transmission interconnection cost and access/service road cost using the same benchmarks it had employed in approving similar projects; and
4. In calculating the tariffs, the ERC used a lower equity internal rate of return ('IRR') of 16.44 percent, except for biomass which was allowed a 17 percent IRR to account for fuel risks.[30]

The ERC also stated that '[t]he approved FITs shall also be subject to review and readjustment by the ERC after the initial FIT implementation of 3 years or when the installation targets for each technology as set by the Department of Energy shall have already been met.'[31] Note the PDOE Resolution Approving the Final Installation Targets, which set the following installation targets: 1. biomass: 250 MW, 2. ocean: 10 MW, 3. run-of-river hydropower: 250 MW, 4. solar PV: 50 MW, and 5. wind: 200 MW.[32] The PDOE's installation targets for solar and wind were lower than those recommended by NREB (which were 100 MW and 220 MW respectively).[33]

ERC Executive Director Francis Saturnino Juan was quoted in the ERC's press release as saying, 'The ERC's lowered FITs will definitely

[29] Energy Regulatory Commission – Philippines, http://www.erc.gov.ph/PressRelease/ViewPressRelease/ERC-Approves-Feed-in-tariff-rates, accessed 9 April 2013.

[30] Ibid.

[31] Ibid.

[32] See Resolution Approving the Feed-In Tariff Rates, supra note 21.

[33] See NREB's Petition to Initiate Rule-making for the Adoption of Feed-In-Tariff, Climate Investment Funds, https://www.climateinvestmentfunds.org/cif/sites/climateinvestmentfunds.org/files/NREB(Petition-FIT).pdf, accessed 9 April 2013.

cushion the impact of implementing the FIT incentive mechanism under the RE Act on the electricity rates, while still being sufficient enough to attract new investments in renewable energy. This is win-win for all.'[34] Certainly, with Filipinos already paying the highest electricity rates in Asia,[35] one can understand the government's heightened sensitivity to imposing higher rates through a tariff system. However, potential project developers will certainly evaluate these tariff rates and their projected IRRs in making investment decisions as to whether to proceed with developing renewable energy projects in the Philippines. In particular, the lowered tariff rates and installation targets for solar and wind may cause concern for investors in those sectors.

The challenge of all feed-in tariff systems is to be able to set the 'right' price to support renewable energy development.[36] Setting the price too low will do little or nothing to develop the market, and setting the price too high will provide excessive profits to developers and impose an unwarranted cost on society. Time will tell whether the ERC got it 'right' (note that the ERC can review and readjust the rates after the initial implementation of three years or when the installation target for the particular technology has been met). The FIT System has the potential to become a 'win-win' arrangement, whereby new project development and financing is attracted to the Philippine renewable energy markets while at the same time not imposing socially inefficient costs on consumers.

D.   CONCLUSION

We can see from the Philippine legal and regulatory framework that the government has an ambitious policy to accelerate the exploration, development and utilization of renewable energy resources, and has recently implemented a FIT System in furtherance of their policy. It should be kept in mind that the Philippines is still a developing country, with a ranking of 112 out of 187 countries in the United Nations Human Development Index and a Gross National Income per capita in Purchasing Power Parity terms of US$3,478.[37] The coming years and decades will test whether the Philippines's renewable energy framework and mechanisms will indeed prove successful in balancing economic growth and development with

---

[34]   Supra note 29.
[35]   Ipsos, supra note 4.
[36]   Cory et al., supra note 16.
[37]   International Human Development Indicators, Philippines, United Nations Development Programme, http://hdrstats.undp.org/en/, accessed 9 April 2013.

reducing dependence on fossil fuels and greenhouse gas emissions, and in moving the country towards the laudable goal of economic and environmental sustainability.

E.   LESSONS LEARNED

The Philippine government has provided very strong incentives for inducing investments in renewable energy, but they are very complex, trying to adjust for all contingencies, and by piling one tax subsidy on top of another, the incentives may be so high as to make renewable energy projects too costly for the government and/or consumers. The very successful German feed-in-tariff is much more simple and, while it may not adjust tariffs for all contingencies, may be more effective.

4. Case study of the implementation of the integrated solar combined cycle pilot plant in Aïn Beni Mathar, Morocco

Richard L. Ottinger with Alexis Thuau

A. INTRODUCTION

Morocco is the only country in Maghreb that does not own important reserves of fossil fuels. In 2007, the country imported no less than 97 percent of its energy needs, being the largest importer in Africa.[1] Thus, Morocco has been pioneering in strengthening its energy security, especially taking the lead in solar and wind energy development.

This study aims at demonstrating the effectiveness of implementation of integrated solar combined cycle (ISCC) for power production using concentrated solar power (CSP) technology in developing countries by addressing the development of a CSP plant located in Aïn Beni Mathar (84 km/52.5 miles to the south of the city of Oujda in the east of Morocco) which was achieved in 2010. CSP plants could efficiently replace centralized plants using fossil fuels. The Aïn Beni Mathar plant serves as an experimental demonstration project before constructing such plants on a larger scale.

Indeed, if development of decentralized renewable energy production (photovoltaics, small individual windmills, etc.) is an important contribution to the improvement of standards of living for people who live far away from the electricity grid, the development of centralized renewable energy power plants linked to the densification of an existing grid is a more relevant response on a longer term. It is the authors' view that the

---

[1] Ministère de l'Energie, des Mines, de l'Eau et de l'Environnement, secteur de l'Energie et des Mines – 'Principales Réalisations' [Major Accomplishments], http://www.mem.gov.ma/Documentation/pdf/PrincipalesRealisations.pdf (accessed: 29 Mar. 2013).

development of centralized energy production which combines a fossil fuel component together with a renewable component, for example Integrated Solar Combined Cycle (ISCC) technology, may be a relevant answer in areas served by a grid to both the development of the electricity distribution in developing countries and the mitigation and adaptation to climate change.

B.   WHAT ARE THE BENEFITS OF ISCC?

- Centralized electricity generation, that allows economies of scale;
- Cheaper costs as compared to decentralized photovoltaic power production;
- Possible implementation in many developing countries;
- Possible storage of solar heat for 12 hours;
- Cheaper, faster and cleaner than nuclear electricity production;
- Likely to be cheaper than fossil-fuel powered plants if subsidies to exploit fossil fuels were cut and the external costs of fossil fuels' combustion were included in the price comparison, and particularly if initial incentives were employed to initiate ISCC projects.[2,3]

C.   PROJECT HISTORY

In November 2009, Morocco announced a plan for the development of five Integrated Solar Energy projects combined with Combined Cycle natural gas units (Projet Marocain de l'Energie Solaire).[4] These projects would account for a total installed capacity of 2,000 MW, equivalent to 38 percent of the installed Moroccan electric capacity in 2009; the estimated investment is expected to reach $9 billion. The five plants are Ouarzazate (500 MW), Aïn Beni Mathar (472 MW), Foum Al Oued (500 MW),

[2]   Ben Sills, 'Fossil Fuel Subsidies Six Times More than Renewable Energy', Bloomberg, 9 Nov. 2011, http://www.bloomberg.com/news/2011-11-09/fossil-fuels-got-more-aid-than-clean-energy-iea.html (accessed: 29 Mar. 2013).

[3]   Natalie Kulichenko, Jens Wirth, 'Regulatory and Financial Incentives for Scaling Up Concentrating Solar Power in Developing Countries', Energy and Mining Sector Board Discussion Paper no. 24, Jun. 2011, http://siteresources.worldbank.org/EXTENERGY2/Resources/CSP_Incentives_DP_24.pdf (accessed: 29 Mar. 2013).

[4]   http://www.masen.org.ma/index.php?Id=42&lang=en#/ (accessed: 29 Mar. 2013).

Boujdour (100 MW) and Sebkhat Tah (500 MW) and are planned to be in operation by the end of 2019. In order to meet these targets, the Moroccan Agency for Solar Energy (MASEN), established in March 2010 as a joint stock company, is in charge of conducting 'overall project design, choice of operators, implementation, management, co-ordination and supervision of other activities related to this program.'[5]

The Aïn Beni Mathar power plant was the first project to be initiated under the public utility Office National de l'Electricité (O.N.E) ownership, although the project was first based on an Independent Power Producer (IPP) concept, which means that the power plant would have been owned by a private company.[6] However, it had to be restructured due to lack of private interest. O.N.E. launched two requests for proposals (RFPs) for ISCC IPP plant projects in May and October 2002, but attracted no interest. Consequently, the project was changed into a public sector project. In 2004, a new RFP was issued, leading to the pre-qualification of four international consortia. In February 2005, bid preparation documents were submitted to the World Bank, and in July 2007, Engineering, Procurement and Construction and Operation and Maintenance contracts (EPC and O&M) were awarded to a Spanish company, Abener (a subsidiary of Abengoa Group), specializing in the development of renewable energy projects, for a duration of five years.[7] After this period, O.N.E. or an IPP would be responsible for the operation of the plant.[8]

An environmental impact study (EIS) was conducted in 2005, ending up with a final report in 2006 which concluded that the project 'will have relatively little impacts on the environment.'[9] Subsequently, a first loan

---

[5] Antonis Tsikalakis, T. Tomtsi, N.D. Hatziargyriou, A. Poullikkas, C. Malamatenios, E. Giakoumelos, O. Cherkaoui Jaouad, A. Chenak, A. Fayek, T. Matar, A. Yasin, 'Review of Best Practices of Solar Electricity Resources Applications in Selected Middle East and North Africa (Mena) Countries', Renewable and Sustainable Energy Reviews, Volume 15, Issue 6, August 2011, pp. 2838–2849.

[6] World Bank, 'Project Appraisal Document on a Proposed Grant from the Global Environmental Facility Trust Fund in the Amount of US$ 43.2 Million to the Office National de l'Electricité of the Kingdom of Morocco for an Integrated Solar Combined Cycle Power Project', 20 Feb. 2007, p. 10, http://www-wds.world-bank.org/external/default/WDSContentServer/WDSP/IB/2007/04/05/000020439_20070405091307/Rendered/PDF/36485.pdf (accessed: 29 Mar. 2013).

[7] 'Morocco Takes the Lead in the Solar Hybrid Race', Modern Power Systems, 1 Aug. 2008, http://www.modernpowersystems.com/features/feature-morocco-takes-the-lead-in-the-solar-hybrid-race/ (accessed: 29 Mar. 2013)..

[8] Ibid. 6, p. 16.

[9] Office National de l'Electricité, Environmental Impact Assessment of the Ain Beni Mathar power plant, http://www.one.org.ma/FR/pdf/eiecentrale.pdf

of €136.45 million ($180 million) was approved in 2005 by the African Development Bank (AfDB) to O.N.E. In 2007, the structural design of the plant was reconsidered, based on the growing demand for electricity. Indeed, the initial project was based on the construction of a 250 MW plant. Therefore, an update of the initial EIS was conducted and issued in March 2007 for a plant with a new capacity of 472 MW.[10] A subsequent €151.40 million ($200 million) loan was granted by the AfDB. Additionally, the Global Environment Facility, through the World Bank, granted €43.2 million ($57 million) to the project, accounting for two thirds of the value of the solar component; the Instituto de Crédito Oficial (ICO), which is the Spanish development bank, approved a loan of €40 million ($53 million) and the O.N.E. itself provided €100 million ($132 million) of its own funds.

The first stone was laid in March 2008, and in May 2010, King Mohammed VI inaugurated the plant, although it took six more months for the plant to become operational.

D.   FUNCTIONING OF THE BENI MATHAR PLANT

The Beni Mathar plant uses a 'combined cycle' of power generation, which consists in the cascade of the two following cycles:

- A gas cycle (Brayton cycle) which exhausts the calories contained in the heat transfer fluid between a maximum temperature and an intermediary temperature which is compatible with the requirements of the next cycle;
- A vapor cycle which exhausts the remaining calories and recycles the losses of the first cycle.

The Beni Mathar plant is composed of two gas turbines (first cycle) which drive two alternators. A generator recovers the heat waste produced by the first cycle and converts it into vapor. The vapor operates a vapor turbine (second cycle) which drives a third alternator. What distinguishes the Beni Mathar plant from other combined-cycle gas plants is that this third alternator is also driven by vapor produced by cylindro-parabolic solar concentrators.

(accessed: 29 Mar. 2013).
[10]   Office National de l'Electricité, Environmental Impact Assessment of the Ain Beni Mathar power plant, http://www.one.org.ma/FR/pdf/majeie.pdf (accessed: 29 Mar. 2013).

The solar component of the power plan relies on a well-proven technology since it has also been implemented in Kramer Junction, California, which is the world's largest solar power plant in activity. This technology, named Solar Electric Generating Systems (SEGS), uses east-west oriented rows of cylindro-parabolic reflectors over a hundred meters long. The reflectors revolve around a horizontal axis in order to follow the course of the sun and concentrate the sun's rays onto a reflectors' focal point.

Then the thermal energy is absorbed by a small iron pipe, inside an in-vacuo glass pipe. The circulating synthetic oil is heated up to 390 °C under high pressure (41 bars). Then the oil is pumped through conventional heat exchangers in order to generate overheated vapor which also drives the abovementioned third generator.

Such an association results in high efficiency (around 56 percent). Therefore, this process reduces the emission of heat wastes by one third as compared to a classic thermal power plant. Both use of a solar component and gain in energy efficiency enable the plant to save around 1,500 tons of CO_2 and 12,000 tons of fuel per year that would be emitted by a classic natural gas power plant.[11,12]

E.   PROJECT SUCCESS FACTORS

1.   Location

The plant is located at the intersection of a high-voltage electricity grid of 225 kV, the Maghreb–Europe Gas Pipeline, and a huge amount of underground water for cooling and cleaning the solar reflectors. In addition, the plant is also located in a region which has little fauna and flora, is sparsely populated, and where agriculture and grazing are the major activities. Finally, the plant is exposed to high solar radiation due to the site's high elevation.[13]

---

[11]   Ibid. 10, p. 23.

[12]   For a more thorough description of the ISCC technology, see Jürgen Dersch, Michael Geyer, Ulf Herrmann, Scott A. Jones, Bruce Kelly, Rainer Kistner, Winfried Ortmanns, Robert Pitz-Paal, Henry Price, 'Trough Integration into Power Plants – A Study on the Performance and Economy of Integrated Solar Combined Cycle Systems', Energy, Volume 29, Issues 5–6, April–May 2004, pp. 947–959 or ESMAP, Review of CSP technologies, 2011, http://www.esmap.org/esmap/sites/esmap.org/files/DocumentLibrary/ESMAP_MENA_Local_Manufacturing_Chapter_1.pdf (accessed: 29 Mar. 2013).

[13]   The German Aerospace Center has developed an advanced Geographical Information System (GIS) driven method to identify possible CSP plant sites in the Middle East–Northern Africa region.

2. Social Benefits

As mentioned by the World Bank:

> The development of the power plant is expected to have a positive social impact on the local population and the local economy a. providing both direct (and indirect) sources of employment during the construction and the production phases (500 direct jobs during the former and 50 during the latter phase); the majority of workers will be locally recruited; b. leading to improvements in local infrastructure including the rehabilitation of local roads; and c. improving access to electricity particularly for poor rural families who are not currently connected.[14]

Under the operation of this new type of plant O.N.E. will have the capacity to train the Moroccan people to build, operate and maintain a hybrid solar combined cycle power plant, which is made of two of the most highly environmentally friendly sources of electricity generation.

3. Environmental Benefits

The Beni Mathar plant requires water in order to clean the solar reflectors and to cool the plant.[15] However, when the decision was taken to raise the plant production capacity from 250 MW to 472 MW, it was also decided not to increase the water consumption but rather use an air cooling system. Therefore the water consumption decreased from 3.1 million cubic meters per year to 1.5 million cubic meters per year, with few consequences on the aquifer.[16]

With regard to air pollution, and as stated in the World Bank project appraisal document, 'the combustion of natural gas does not entail SO2 emissions, and has significantly lower emissions of NO2 than other fossil fuels.'[17] The use of a solar component also reduces the emissions of greenhouse gases (GHGs) by about 1,500 tons per year as compared to the operation of a classic natural gas plant.

Soil pollution also has been taken into account. The major concern in this regard may arise from accidental contamination of the site by a spill of synthetic oil during the operation of the plant. According to the EIS and its update, an emergency plan is required in order to manage the potential risk of accidental discharge of hydrocarbons or synthetic oil, including information on the way these pollutants should be treated, adequate

---

[14] Ibid. 6, p. 26.
[15] Ibid. 6, p. 41.
[16] Ibid. 10, p. 8.
[17] Ibid. 6, p. 27.

procedures in case such an incident occurs in an urban, rural or on-site context and detailed proposals for staff training.[18]

4.   Capacity Building Enhancement

Lessons can be learned from the implementation of the Beni Mathar plant in order to replicate the project in a more efficient fashion. As a matter of fact, enhancement of capacity-building is a condition laid down by the World Bank for the issuance of its €43.2 million ($57 million) grant; the World Bank required the O.N.E. to 'disseminate the results from the project both domestically and internationally, as a way to support future replication.' The World Bank further stated that 'O.N.E. will also make all the information about the ISCC power plant, including performance indicators, available on its external website.' Moreover, information sharing must also be made 'through facilitating visits by the power industry, utilities and other interested institutions from all over the world to learn from the construction and operation of the plant and the results achieved.'[19,20]

5.   Replicability

By implementing the Beni Mathar plant, O.N.E. gained experience in terms of design, construction, operation, training of personnel, etc. Since the project was implemented successfully so far, there are chances that O.N.E. will use the acquired knowledge to replicate the technology elsewhere in Morocco or in the Middle East and North Africa (MENA) region. The more the technology spreads, the more economies of scale will be realized. Consequently, the costs will decline to commercially competitive levels. This experience should be of great benefit to other countries choosing to adopt this technology.

F.   OBSTACLES ENCOUNTERED

1.   General Risks in Investing in a Developing Country

Investors are usually reluctant to engage in investments that may endanger the security of their personnel or alter the operation of their installations.

---

18   Ibid. 9, p. 104.
19   Ibid. 6, p. 16.
20   Ibid. 10, p. 41.

In this regard, Morocco is likely to attract foreign investments, due to its rather stable political structure.

Investors also are reluctant to finance unproven technologies in developing countries. Thus, through the operation of the Beni Mathar plant, Morocco has been implementing ISCC technology for the first time on its soil, whereas no single statute or regulation applies to such an installation. The EIS consequently calls for the implementation of a legal framework to protect investors, stating that 'the development of this new type of energy shall be inserted in the statute 11-03 [relating to the protection and valuation of the environment].' If these provisions are inadequate, developers would be facing legal uncertainty and feel that their investments were not safe.

One other issue lies in the lack of economic incentives to engage in a renewable energy project. These projects often are costly because the people who need them the most, poor people who live in rural areas, lack the financial resources to pay for them. Also, the use of fossil-fueled power plants is made easier because developing countries often encourage the exploitation of their fossil fuel resources, such as coal, oil and gas, through the use of subsidies. However, Morocco relies heavily on fossil fuel imports. Thus, oil represented 22.5 percent of its total imports in 2010, which gave the country incentives to encourage local, renewable energy production, with the help of donors such as the Islamic Development Bank, the European Investment Bank and the World Bank.[21] Therefore, strong support was given to developers of such alternative energy technologies.

2. Strengths and Weaknesses Inherent to the Beni Mathar Project

Social acceptance of the project was a major concern as the development of renewable energy often is not considered a promising and safe investment for developing countries. In the present case, the expropriation process for the 29 families whose land and/or assets would be affected by the construction of the power plant and the gas pipeline provides a good illustration of the steps taken to promote acceptance of the project:

- The land concerned with the expropriation process was essentially used for crops and it was easy for the farmers to resettle these crops elsewhere due to low pressure on land;

---

[21] Energy Sector Management Assistance Program (ESMAP), 'Power Sector Financial Vulnerability Assessment – Impact of the Credit Crisis on Investments in the Power Sector: The Case of Morocco', Jan. 2011, p. 14, http://www.esmap.org/esmap/sites/esmap.org/files/Esmap%20Vulnerability%20Morocco%2001%2012.pdf (accessed: 29 Mar. 2013).

- The relatively small number of persons affected facilitated the granting of fair compensation;
- In October 2005, informal consultations were conducted with pastoralists in the immediate vicinity of the proposed plant.[22] Then, a social assessment was conducted in 2006 through 'a series of intensive consultations', which confirmed that 'the populations are awaiting the employment opportunities which the project will likely offer.'[23]

Also, a public consultation was held in December 2005 in the Municipality of Aïn Beni Mathar to raise awareness among the invited participants and to address their concerns.[24]

The project's social acceptance enabled the O.N.E. to make use of the African Development Bank guidelines, which enable the developer to take 'anticipated actions,' that is, before the project's approval by the AfDB's Board of Directors is deemed applicable it must be clearly demonstrated that anticipated signature of contracts for the acquisition of land will be a crucial factor in the timely implementation of the project.[25]

The Energy Sector Management Assistance Program (ESMAP) conducted an evaluation of the project's achievements at the end of 2010 which stated the following strengths and weaknesses of the pre-operation phase:[26]

- All main components and equipment are imported for the Aïn Beni Mathar project from international market players;
- International EPC contractor Abener commissioned the project successfully by the end of 2016 (duration of construction will be 3 years);
- Abengoa and Abener observed small cost differences for metal mounting structures in Morocco because of the small margin

---

[22]   Ibid. 6, p. 64.
[23]   Ibid. 6, p. 62.
[24]   Ibid. 6, p. 7.
[25]   See African Development Bank, 'Projet de développement du réseau de transport et de répartition d'électricité' [Electricity transmission and distribution development project], 2009, p. 8, fhttp://www.afdb.org/fileadmin/uploads/afdb/Documents/Project-and-Operations/Maroc%20-%20Rapports%20d'évaluation.pd (accessed: 29 Mar. 2013).
[26]   ESMAP, 'Review of Manufacturing Capabilities and Potential in MENA Countries', p. 93, http://www.esmap.org/esmap/sites/esmap.org/files/DocumentLibrary/ESMAP_MENA_Local_Manufacturing_Chapter_2.pdf (accessed: 29 Mar. 2013).

between imports and local manufacturers (no advantage for local components);

• Abengoa and Abener lacked significant international experience in CSP, which made contracting with local companies more complex.

• There were doubts about the ability of local industry to supply the necessary equipment and technical support in adequate quantity and in a timely fashion, and therefore it was considered a lower risk to buy from large international suppliers;

• Problems of finding well-trained and highly skilled workers;

• Problems of local products and steel construction: mainly quality and price;

• Issues related to intra-Morocco logistics: importing components from abroad seemed easier than shipping by road from economic hubs (Casablanca, for example) to Aïn Beni Mathar;

• Unavailability in Morocco of large machines (all types);

• Administration and bureaucracy impediments resulting in a lower speed of implementation of the project;

• Reasons for imports: no import taxes resulted in price disadvantages for local products (a price difference of 6–10 percent);

• O&M had to be undertaken by internationally experienced companies to sustain the performance of the plant.

The study thus concluded that 'local manufacturing outcomes have not proved positive' in terms of participation of the local industry for the implementation of the project and reliability of local components and services.

It must also be noted, in relation to the ESMAP comment on the lack of well-trained and highly skilled workers, that inadequate safety rules might also have been an important area of consideration. Indeed, on May 22, 2009, a fire started which partly damaged components of the plant's gas turbine no. 2. A human factor might have been involved (a thoughtlessly discarded cigarette, for example), which stresses the need to brief workers more extensively on safety rules, especially in locations subject to specific hazards.[27]

---

[27] 'Aïn Béni Mathar: Incendie à la centrale thermosolaire' [fire from concentrated solar thermal power installations], La Gazette du Maroc, 29 May 2009, http://www.lagazettedumaroc.com/articles.php?id_artl=20436 (accessed: 29 Mar. 2013).

G.   CONCLUSION

The implementation of the Beni Mathar project has paved the way for the implementation of subsequent projects. Indeed, it has provided Morocco with a better understanding of the ISCC technology and benefits in terms of environmental sustainability and developing a global partnership for development (objectives 7A[28] and 8F[29] of the Millennium Development Goals[30]).

Whereas the World Bank's appraisal document for its $43.2 million grant reads: 'a mid-term review will be carried out after the first year of operation to evaluate the achievement of the project development objective',[31] no evaluation of the implementation of the Beni Mathar plant has been issued yet. The following indicators may be considered in conducting the evaluation:

- Quantitative indicators such as the relative cost of solar thermal as compared to the cost of natural gas and other power production methodologies, the reduction in CO_2 emissions achieved.
- Qualitative indicators such as the framing of an optimal bidding process, the extent to which capacity-building and the CSP technology can be transferred to other countries, the transparency of the plant's operation so as to enable its dissemination, the real figures of local jobs creation, the extent to which such a technology benefits the population based on several criteria (health improvement, poverty alleviation, etc.), the impacts on the environment, the lessons learned from the pre-operation, construction and operation phases.

Many consortiums and companies have expressed a great interest in the development of solar thermal plants, as MASEN's latest RFP in relation to the Ouarzazate plant showed. Indeed, this project was warmly welcomed worldwide, with around 200 answers to the RFP.[32] Developing

---

[28]   Objective 7A reads: 'Integrate the principles of sustainable development into country policies and programs; reverse loss of environmental resources.'

[29]   Objective 8F reads: 'In co-operation with the private sector, make available the benefits of new technologies, especially information and communications.'

[30]   Available at United Nations, http://www.un.org/millenniumgoals/ (accessed: 29 Mar. 2013).

[31]   Ibid. 6, p. 16.

[32]   Silvia Pariente-David, World Bank, 'CTF-Ouarzazate I CSP Project', 16 Oct.   2011,   http://climatepolicyinitiative.org/event/wp-content/uploads/2011/12/

countries should therefore be made aware of such interest expressed by investors.

H. LESSONS LEARNED

A particular reason for including Morocco's renewable energy experience in the book is its demonstration of the courage and skill to successfully implement a promising new technology, an integrated solar combined cycle (ISCC) power plant utilizing concentrated solar power with natural gas.

The principal advantages of the ISCC technology are that the intermittency of the solar power is ameliorated by the combined use of natural gas, and the natural gas plant greenhouse gas and other natural gas plant pollutants are mitigated by the combined use of solar energy. The result is a reduction of about 1,500 tons per year of greenhouse emissions from a solely natural gas power plant.

ISCC is also considerably cheaper than importing and utilizing fossil fuels, particularly if their external costs are included.

Full consultation was held with the 29 families who would be displaced by the project and their consent was obtained. Comprehensive public hearings also were held with favorable results. This social acceptance facilitated international bank loan approvals.

Morocco has plans for five ISCC plants with a combined 2,000 MW capacity representing 38 percent of the country's power requirements. The ISCC technology is replicable in other countries, and the international lenders required that all aspects of the technology be publicly disclosed and that the plants be made available for inspection by other countries. A great deal of interest was expressed as indicated by the approximately 200 responses received to the RFP for the Ouarzazate plant.

The difficulties encountered included investor reluctance to financing of a new technology. This problem was overcome by persuading the African National Bank and the World Bank of the soundness of the project, facilitated by the fact that stand-alone concentrated solar and natural gas power are proven technologies. Also Morocco was willing to comply with the stringent social requirements of the AfDB's lending conditions.

Another obstacle was that Morocco had no renewable energy statute setting forth the legal requirements of projects and the legal responsibility

Pariente-David_CTF-Ouarzazate-I-CSP-Project.pdf (accessed: 29 Mar. 2013).

of developers. This deficiency has since been addressed generally and the terms and conditions of the international loans set forth the requirements for the original plant.

Lastly, an evaluation by the Energy Sector Management Assistance Program (ESMAP) of the project's strengths and weaknesses as of the end of 2010 noted weaknesses in the country's ability to initiate and manage such projects on its own because of insufficiency of equipment that had to be entirely imported, the inadequate experience of the Spanish company that Morocco selected to construct the initial project, the insufficient number of trained skilled local technicians and administrative impediments in the local government.

However, the implementation of the initial project provided an important learning experience for the Moroccan government and it has taken action to address the problems revealed. The demonstrated strengths of the ISCC technology in terms of energy reliability, environmental advantages and cost are sufficient to persuade the government to continue the remainder of its planned domestic program and respond to the interest expressed in the ISCC technology from many other countries.

5.   Case study of biofuels in India

Richard L. Ottinger with Sayan S. Das

PART I

India biodiesel and bioethanol

A.   INTRODUCTION

Biofuels are fuels derived from biomass – from plants and other naturally occurring organic materials. They are renewable energy sources, unlike finite energy resources such as petroleum, coal, and nuclear fuels.[1]

Biofuels have witnessed a surge in production and use the world over in recent years. Their attractiveness is many-fold if employed with adequate environmental and social protections:

1. They can enhance energy security by relieving dependence on depleting and insecure oil supplies from unstable or hostile countries;
2. They can reduce emissions of greenhouse gasses and other pollutants;
3. They can create employment and save countries, businesses and individuals money;
4. They may be the only affordable resources available to developing countries to promote their economic development since biofuels can use crops and waste products readily available or growable locally in most developing countries, and they do not require expensive or complex refining equipment or expertise importation;
5. They can relieve the burden on women and children from gathering firewood and using the wood for cooking, often under asphyxiating conditions;
6. They can provide affordable electricity for night reading, education, communications, and refrigeration for food and medicines.

---

[1]   'What is Biofuel?,' http://biofuel.org.uk/what-are-biofuels.html (last visited 13 February 2012).

But the qualifications that biofuels be deployed only with adequate environmental and social protections are vital. These include:

1. Use only of feedstocks that are sustainable:

    a. That do not involve deforestation;
    b. That do not involve food crop displacement or result in food short-ages or price increases;
    c. That do not involve monocultures that over time will render land unusable;
    d. That do not create water or air pollution;
    e. That do create positive greenhouse gas emissions reduction; and
    f. That minimize the use of chemical fertilizers and pesticides, use as many biodegradable materials as practicable and manage waste safely; and
    g. That avoid or control invasive species.

2. Give incentives for use of biofuels grown in poor and rural areas;
3. Provide for public participation in decision-making for all projects;
4. Require that an environmental assessment be made of all projects to identify potential threats;
5. Provide for adequate training of project participants;
6. Provide for contracts with providers to maintain equipment and supply spare parts;
7. Observe all local and international labor laws and provide safety measures for workers and medical attention for anyone injured; and
8. Where feasible, obtain a certificate of sustainability from the Roundtable for Sustainable Biofuels (RSB).

With regulations adopted to control these environmental and social risks, biofuels have the potential to be a very large and significant clean energy resource. This is particularly true as present and future biofuel feedstocks, such as those from algae and various waste products, are proven feasible.

Brazil, the United States and the European Union countries are leaders in production, use and export of biofuels, with Brazil exporting 5 billion liters of ethanol fuel in 2008.[2] These countries have taken some steps to

---

[2] Stephan Bringezu et al., 'Towards Sustainable Production and Use of Resources: Assessing Biofuels,' UNEP, 2009, http://www.unep.fr/scp/rpanel/pdf/

adopt the safeguards recommended, though none of them have adopted all and only a very few (none of the above) have yet sought to obtain RSB certification.

Biofuels in India are mainly derived from oil rich Jatropha plants, Jatropha curcas. These plants have exhibited many sustainable qualities. While other plants can grow on wasteland, eliminating the need to replace production of other crops from active farmlands, Jatropha unfortunately thrives better in good soil. However, the plants can be spread out and food crops can be intercropped with the Jatropha. The plant requires little fertilizer when grown on fertile land and also few pesticides. It is carbon neutral and oil can be extracted from it without the added requirement of refining, though it usually is refined to produce biodiesel.[3]

India's crude oil and petroleum product supplies are largely import-dependent. With oil import expenditures increasing by more than six times in the last 25 years due to escalation in global demand and price increases, the demand for biofuel is expected to increase significantly in coming years. Biofuels will be critical in reducing dependence on fossil fuels, achieving greater energy security, and reducing noxious emissions.[4]

With incentives, it is only natural for India to lean towards the production of biofuels, and particularly Jatropha, to produce bio-diesel which could easily replace the diesel that fuels most of India's transportation.[5]

India produced roughly 880,000 barrels per day (bbl/d) of total oil in 2009 from 3600 operating oil wells. Approximately 680,000 bbl/d was crude oil; the remainder was other liquids and refinery gains. In 2009 India consumed nearly 3 million bbl/d of oil, making it the fourth largest consumer of oil in the world. The US Energy Information Agency (EIA) expects approximately 100,000 bbl/d annual consumption growth through 2011. The experts at the Center for Jatropha Promotion and Biodiesel (CJP) have stressed that there is an urgent need to search for an option to substitute for this increase, claiming that the end of oil availability is not very far off. They have suggested that biofuels can substitute for this need for additional oil, and that the Indian government should adopt a concrete

---

assessing_biofuels_full_report.pdf\ (last visited 13 February 2012).

[3]   Michael Fitzgerald, 'India's Big Plans for Biodiesel,' http://www.technolo-gyreview.com/Energy/17940/ (last visited 13 February 2012).

[4]   Indian Bio Fuels Market, http://www.fuerteventuradigital.com/notias/News/2007/09/19/194531.asp (last visited 13 February 2012).

[5]   India consumed 4.96 million tons of diesel in October 2010 as opposed to 1.21 million tons of petrol, http://www.domainb.com/industry/oil_gas/20101203_india_diesel.html (last visited 13 February 2013).

biofuel program and strategy to provide it with energy security as well as national security.

In this context CJP is targeting to plant 20 billion biodiesel producing trees by 2013. At present India imports around 800 million bbl crude oil per annum. To overcome the anticipated deficit India would need to import 8 million bbls of additional oil, requiring an investment of US$ 4 trillion. To avoid these impacts, a plantation of 20 billion biodiesel producing trees can be planted with an investment of US$ 40 billion, potentially producing 900 billion bbl of biodiesel per annum with amazing other benefits to the environment, increased employment and rural electrification. The present proposed project explores a method of growing oil seed trees in nontraditional agronomic areas. It would generate the following positive outcomes:

- Production of sustainable fuels in ideal areas that are not costly to maintain
- Carbon reductions
- Support to the local economy by providing jobs
- Relatively inexpensive fuel[6]

India has 68.35 million hectares of wasteland of which 50 per cent is non-forest land that can be made fertile again.[7] With such a bounty of land, this is an opportunity to foray into Jatropha cultivation with success without having to alter agricultural production.

The cost of manufacturing biodiesel in India will be about Rs. 21 or US$.40 cents per liter.[8] Petroleum prices in India are on the rise, recently taking the price of a liter of petrol to Rs. 66.84 or approximately US$1.30.[9] With such high rates and inflation rates reaching a record high of 9.78 per cent,[10] prudence demands biofuels to be employed with enthusiasm.

---

[6]   Center for Jatropha Promotion and Bio Diesel, http://www.Jatrophaworld.org/ (last visited 12 February 2012).

[7]   Integrated Wasteland Development Programme, Ministry of Rural Development, Govt. of India, http://dolr.nic.in/iwdp1.htm (last visited 12 February 2012).

[8]   'An Assessment of the Biofuels Industry in India,' http://www.unctad.org/en/docs/ditcted20066_en.pdf (last visited 13 February 2012).

[9]   Petrol prices are predicted to rise again and another hike is said to be in the offing, http://www.livemint.com/2011/09/15175541/Petrol-prices-rise-again-anot.html (last visited 13 February 2012).

[10]   'India Raises Interest Rates to Fight Inflation,' http://www.bbc.co.uk/news/world-south-asia-14942749 (last visited 12 February 2012).

B. PROJECTS

1. Government of India

The Government of India has formulated the National Biodiesel Mission to meet 20 per cent of the nation's energy requirements by 2011–12.[11] This is a big step forward by the Government in its hope to diversify its energy sources. The policy aims to mainstream the use of biofuels across every usage possible starting with transportation. The broader aim is to find effective substitutes for fossil fuels. An indicative target of 20 per cent blending of biofuels both for biodiesel and bioethanol by 2017 is proposed.

The focus for development of biofuels in India will be to utilize waste and degraded forest and non-forest lands only for cultivation of shrubs and trees bearing non-edible oil seeds for production of biodiesel. Cultivators, farmers, landless laborers, etc. will be encouraged to undertake plantations that provide the feedstock for biodiesel and bioethanol fuels. Corporations will also be enabled to undertake plantations through contract farming involving farmers, co-operatives and self-help groups. Appropriate financial and fiscal measures will be considered from time to time to support the development and promotion of biofuels and their utilization in different sectors. Emphasis will also be given to the development of second generation biofuels and other new feedstocks for the production of biodiesel and bioethanol.

Some of the projects contemplated are described below.

2. Chhattisgarh

Both public and private sector companies have approached the Chhattisgarh government seeking permission for contract farming of Jatropha. The state government is likely to make available about 20 lakh or 2 million hectares of land for cultivation.

Under the Government scheme, an individual can lease up to 200 hectares of land at a price of Rs. 100 or US$2 per hectare, per year for the first five years. For subsequent years, these rates could be increased. The Chhattisgarh government has estimated the cost of production at Rs. 20 per liter if the cost of seeds is taken at Rs. 5 per kg. This corresponds to the landed cost of fossil diesel at US$45 per barrel. With duty concessions, biodiesel proves very competitive at US$35 a barrel.

---

[11] National Policy on Biofuels, Government of India, Ministry of New and Renewable Energy, http://www.svlele.com/nbp.pdf (last visited 12 February 2012).

The state government has already formulated an action plan including the setting up of the Chhattisgarh Biofuel Development Authority, identifying Government owned waste or fallow land as well as constituting task forces in various districts. The action plan envisages encouraging the private sector to invest in contract farming, setting up oil expeller and transesterification plants[12] for biodiesel production in Raipur and smaller units to be set up in every other district or county (US equivalent).[13]

3. State Bank of India

In 2010, India's largest national bank, the State Bank of India, entered into an agreement with D1 Mohan Bio Oils Ltd., an Indo-UK business partnership, to give Rs. 130 crores or US$56 billion (approximately) as loans to farmers in Southern India to cultivate Jatropha seeds on 100,000 acres of land.[14]

C. ADVANTAGES OF BIOFUELS

The general advantages of biofuels in India are fourfold:

- They will reduce greenhouse gas emissions;
- They will allow large tracts of wasteland to be brought under cultivation;
- With our huge human resource, cultivation of these lands could turn into major employment resources; and
- India could gain vital energy security.

---

[12] Transesterification is the stage in which raw oil is transesterified to biodiesel, which is methyl or ethyl ester based on whether methanol or ethanol is used in the production process. The capacity of the transesterification plant is dependent on the amount of raw oil that has to be transesterified into biodiesel. The capital cost of the transesterification plant depends on its capacity. The process of oil extraction has been existent from very early times. In those days, the seeds were merely cleaned and ground with the use of grinders or stone mills. With the advancement of science and technology, the use of machines like screw presses or oil expellers was started for expelling or extracting oil from oil seeds and nuts. Oil expellers play a very vital role in the oil extraction process.
[13] 'India Inc Eye Jatropha Plantation,' http://www.thehindubusinessline.in/2005/09/02/stories/2005090202950100.htm (last visited 12 February 2012).
[14] 'SBI to Provide Loan for Jatropha Cultivation in TN,' http://www.thehindubusinessline.com/todays-paper/tp-agri-biz-and-commodity/article2174262.ece (last visited 12 February 2012).

D. ISSUES

The lack of large-scale availability of feedstock restrains the market. Biodiesel will take a while to establish itself as an effective biofuel, since Jatropha plantations in the country are still in the initial stages of development and it takes several years for plants to mature sufficiently to produce significant amounts of oil. Three to four years and many plantations later, the country may have the feedstock necessary for the large-scale production of Jatropha oil for use in biodiesel.

The absence of a clear Government policy on biofuels and lack of availability of domestic feedstock has inhibited several biofuel manufacturers from entering this market. Hence, Indian manufacturers are considering importing palm oil to produce biodiesel.

Another issue with the manufacture of biodiesel in India is the per hectare income from Jatropha cultivation is not as high as farmers would like, also seed collection and oil extraction infrastructure is lacking as is the method of utilizing glycerol, the by-product of the extraction, which accounts for 12 per cent of the extraction.[15] As the industry is in its infancy these issues will be part of the building process.

The main problem in getting the biodiesel program rolling has been the difficulty in initiating the large-scale cultivation of Jatropha. Farmers do not consider Jatropha cultivation is rewarding enough. To alleviate this problem, confidence-building measures need to be taken. The Government should clearly formulate its policy and explain to the farmers that their role is of vital importance to the success of the biodiesel program. The Government also needs to establish a minimum support price for Jatropha oil seeds to assure farmers of adequate and timely payments.

The other main issue is the lack of infrastructure in seed collection and oil extraction. In the absence of infrastructure and available oil seeds, it will be difficult to persuade entrepreneurs to invest in transesterification plants.

Finally, there is the problem of glycerol utilization. The by-product of glycerol is about 12 per cent of the biodiesel produced, and is of about 88 per cent purity. If no alternative means is quickly devised for utilizing glycerol, its price will plummet due to excess supply.

Similar availability issues also affect the better-developed Indian bioethanol industry, as ethanol is primarily manufactured in India from molasses, a by-product of sugar. Since sugarcane production is cyclical, the

---

[15] 'An Assessment of the Biofuels Industry in India,' http://www.unctad.org/en/docs/ditcted20066_en.pdf (last visited 12 February 2012).

availability and cost of production of bioethanol will vary depending on sugarcane crop yields. India's ethanol blending program could not be implemented during 2003-04 due to a low sugarcane output, and the second phase of this program was announced in September 2006 only after a recovery in sugarcane production.[16]

The Government needs to reform restrictive policies to loosen constraints on ethanol production. For instance, a ban on cross-state movement of molasses should be removed. Ethanol distilleries should be allowed to use sugarcane juice instead of just molasses for ethanol manufacture. When sugar prices are depressed, this would permit sugarcane farmers to divert some of the sugarcane to ethanol production, thus bringing extra income to the farmers.

In addition, state and central government policies should be harmonized. State governments stand to gain windfall revenue from an excise tax on potable ethanol and would hence prefer a substantial portion of the ethanol production to be earmarked for potable use.

The central government is charged with energy policy and the procurement of ethanol for blending. The wide fluctuations in the price of molasses, which is the main determining factor in the cost of ethanol, should be brought under control. In order to protect itself from the volatilities of molasses prices, the alcohol-based industry is demanding that futures trading be allowed for the commodity. This would take away the spikes in the prices, as well as smooth prices to more realistic levels.[17]

PART II

India's biogas program

A.   INTRODUCTION

Mahatma Gandhi had stated that real India resided in its villages. In these villages he sought to implement a system which was sustainable in its nature and made the immediate community self-sufficient in their needs while creating an environment of co-operative economic structures.[18] Keeping this vision in mind, biogas in India is pivotal for the growth of the

---

[16]   Ibid.

[17]   Ibid.

[18]   Jo Lawbuary, 'Biogas in India: More than Gandhi's Dream?,' http://www.ganesha.co.uk/Articles/Biogas%20Technology%20in%20India.htm, (last visited 2 May 2012).

country. With the livestock population in India being 529.7 million[19] and the yearly dung production exceeding 1500 million tons,[20] biogas could be a great source to bring better social and economic parity to rural India while inculcating a lifestyle based on sustainability and sound energy creation mechanisms. Biogas is an astonishing tool to bring about a new era of green and sustainable energy in rural India. This will also bring in new social and economic order along with greater job creation.

Manure, the principal biogas feedstock, is readily available, mostly at no cost. This gives the Government a potent tool to tackle issues of providing an affordable fuel for rural communities. Moreover providing organic manure for households through family biogas plants reduces pressure on forests for firewood and subsequent pollution, relieving women and children from the burden of gathering firewood and the danger from pollution in using it for cooking.[21]

Programs related to biogas in India are managed by the Ministry of New and Renewable Energy (MNRE) under the leadership of Dr. Farooq Abdullah. It is an independent central ministry charged with all forms of renewable energy in India. The Indian government has been committed to establishing biogas infrastructure, both commercial and non-commercial, as it saw biogas as an essential tool to diminish the country's dependence on expensive and polluting conventional fuels. Though its expansion came on the heels of wrongful data that fuel wood was the main cause of deforestation in the 1970s, when in reality industrial and urban use were the principal cause, the result of this expansion has held rural India in good stead. Since 1981–82 the Government of India has been implementing the National Biogas and Manure Management Programme (NBMMP) which has resulted in about 4.31 million household and community biogas plants across India.

A biogas plant uses organic waste material, primarily animal excreta, to convert into gas. This gas consists of methane and carbon dioxide. The conversion of manure into gas results in a by-product which can be used as organic fertilizer. This by-product is called slurry. A family size biogas

---

[19] 'Livestock Population in India by Species,' National Dairy Development Board, http://www.nddb.coop/English/Statistics/Pages/Population-India-Species. aspx (last visited 2 May 2012).
[20] 'The Importance of Cattle in Biogas Production,' http://www.preservearticles.com/2012042130934/the-importance-of-cattle-in-biogas-production.html (last visited 2 May 2012).
[21] 'Probable Questions for Biogas Programme,' Ministry of New and Renewable Energy, http://mnre.gov.in/file-manager/UserFiles/faq_biogas.htm (last visited 2 May 2012).

plant is the most common form found in India. The Government of India estimates that there are in India about 12 million family type biogas plants. This statistic has been arrived at after estimating the availability of cattle dung in India.[22]

B. HISTORY

India was one of the pioneering countries, using biogas as far back as the 1920s. The Indian Agricultural Research Institute was the first such institution to start research on biogas, and it was followed by independent research in centers like that in Mumbai (then Bombay). The initial attraction was the manure's use as fertilizer and use as a fuel was considered a by-product.[23] In 1938 pilot biogas programs were started by the British Indian government. The first such unit was developed at the Mantunga Homeless Lepers Asylum near Mumbai.[24]

One of the earliest units was christened Grama Lakshmi, which means the rural goddess of wealth. This unit had a digestion chamber which was built underground, a gasholder and a reactor all in the same unit to save material and space, a continuous flow system, automatic overflow when loading and a scum breaker to prevent scum from entering the gas pipe.[25] In 1952 these innovations were demonstrated for practical use by the Khadi and Village Industries Commission (KVIC) but none of them bore fruit. The KVIC then developed the floating drum design.

In 1961 the state sector adapted the Chinese dome design as a Janata plant (4 to 10 m³), which later in 1981 was adapted as the Deenbhandu plant (2 to 8 m3), with the Government giving a subsidy for such designs. With the oil crisis in the 1970s, the Government was forced to look for alternatives to fossil fuels, and thus commissioned 50,000 biogas plants of which 70 per cent were built.

President Indira Gandhi introduced subsidies for biogas installation to promote energy independence, alleviate poverty, and because it required skill and considerable cost for installation thus without subsidies, relegated it to well-to-do households. Thus subsidizing took biogas from the hands

---

[22] Mathias Gustavsson, 'Biogas Technology – Solution in Search of its Problems,' Gotenborg University, 2000, http://www.ted-biogas.org/assets/download/Gustavsson2000.pdf (last visited 2 May 2012).

[23] Ibid.

[24] 'A Brief History of Biogas,' University of Adelaide, http://www.adelaide.edu.au/biogas/history/ (last visited 2 May 2012).

[25] See Gustavsson, supra note 22.

of local research groups to national research centers and subsequently the politicians were making decisions on biofuels. By the end of the decade the All India Coordinated Biogas Program was launched which tackled deforestation and the expense of oil imports.[26] But at that time the technology of biogas was accessible primarily to the rich and educated farmers, and so the movement withered away. In 1997, the Tata Energy Research Institute carried out research which found 80 per cent of the biogas plants were performing below expectations due to technical difficulties. In 2009, India recorded about 12 million plants with a success rate of 60 per cent.[27]

The impetus to implement household biogas plants to a broader economic base began in India only in 1981 with the Government of India implementing the National Project on Biogas Development (NPBD), after which biogas was featured in the 20 point National Program and became a national priority.[28] The program used the 'Multi-Model' and 'Multi-Agency' approach wherein non-governmental organizations (NGOs) played a central role in achieving the construction of about 3 million household biogas plants, which has also ensured wider coverage throughout the country. It is still a drop in the ocean when considering the revised potential (in year 2000) of 20 million household plants. At the present rate of implementation of 150,000 units per year, it could take 75–100 years to realize this potential.[29]

In the Ninth Five Year Plan (1997–2000), it was decided that direct subsidies to families to install biogas plants would be phased out and to make biogas commercially viable. At the same time, it was decided that decentralization of implementation would take place which would make implementation a part of the state governments' administrative duties, away from the hands of the Ministry.[30] This project was later rechristened as the National Biogas and Manure Management Programme.[31]

---

[26]   Ibid.
[27]   Dr. David Fulford, 'Biogas Technology: Successful Projects in Asia and Africa,' Kingdom BioEnergy, http://www.kingdombio.com/DJF%20Biogas%20 Reading%20presentation.pdf (last visited 2 May 2012).
[28]   See Gustavsson, supra note 22.
[29]   Raymond Myles, 'Implementation of Household Biogas Plants by NGOs in India,' http://www.inseda.org/Additional%20material/Lessons%20learnt%20 NGOs%20Biogas%20program.pdf (last visited 2 May 2012).
[30]   See Gustavsson, supra note 22.
[31]   The National Biogas and Manure Management Programme (NBMMP), http://www.indg.in/rural-energy/schemes/national-biogas-and-manure-manage ment-program-nbmmp (last visited 2 May 2012).

C.   GOVERNMENT POLICIES

1.   The National Biogas and Manure Management Programme (NBMMP)

The National Biogas and Manure Management Programme aspires to provide gaseous fuel for cooking, lighting and enriched organic fertilizer as a by-product, besides as a type of waste disposal system at the domestic level.[32] Funded by the central government, this project provides financial assistance for installation of biogas plants, turn-key job fees, training, technical support and publicity.[33]

The program is implemented by state nodal departments and agencies and the Khadi and Village Industries Commission (KVIC), Mumbai. APITCO Ltd. Hyderabad conducted an independent evaluation on the biogas plants installed in the Tenth Five Year Plan (2002–07) in six states across India. The report was submitted in 2009 which stated that 95.81 per cent of the plants were working.[34]

A total of 4.31 million family biogas plants have been installed under this program since its inception.[35] A target of setting up of 6.47 lakh family type biogas plants has been fixed with a planned outlay of Rs. 562.00 crore for the Eleventh Plan (2007–11).[36]

The Ministry of New and Renewable Energy has data on the number of biogas plants.[37]

The objective of this program is:[38]

---

[32]   Initiatives in Good Governance, Ministry of New and Renewable Energy, http://mnre.gov.in/file-manager/UserFiles/Initiatives_in_Good_Governance.pdf (last visited 2 May 2012).

[33]   Details of the National Biogas and Manure Management Programme (NBMMP), http://india.gov.in/govt/viewscheme.php?schemeid=276 (last visited 2 May 2012).

[34]   The National Biogas and Manure Management Programme (NBMMP), Ministry of New and Renewable Energy, http://www.mnre.gov.in/schemes/decentralized-systems/schems-2/ (last visited 2 May 2012).

[35]   'Probable Questions for Biogas Programme,' Ministry of New and Renewable Energy, http://mnre.gov.in/file-manager/UserFiles/faq_biogas.htm (last visited 2 May 2012).

[36]   The National Biogas and Manure Management Programme (NBMMP), Ministry of New and Renewable Energy, http:/www.mnre.gov.in/schemes/decentralized-systems/schems-2/ (last visited 2 May 2012).

[37]   'Achievements,' Ministry of New and Renewable Energy, http://www.mnre.gov.in/mission-and-vision-2/achievements/ (last visited 29 April 2012).

[38]   Objectives of the National Biogas and Manure Management Programme

- To provide fuel for cooking purposes and organic manure to rural households through family type biogas plants;
- To mitigate drudgery of rural women, reduce pressure on forests and accentuate social benefits; and
- To improve sanitation in villages by linking sanitary toilets with biogas plants.

The components of this program are:[39]

- Indigenously developed models of biogas plants are promoted;
- States have designated nodal departments and nodal agencies for implementation. Besides the Khadi and Village Industries Commission, Mumbai, the National Dairy Development Board, Anand (Gujarat), and national and regional level non-governmental organizations are involved in implementation;
- The project provides for different types of financial incentives including central subsidy to users, turn-key job fee to entrepreneurs, service charges to state nodal departments/agencies and support for training and publicity;
- Various kinds of training programs are supported. Biogas development and training centers, functioning in nine major states, provide technical and training backup to state nodal departments and nodal agencies; and
- Commercial and co-operative banks provide loans for the setting up of biogas plants in Agricultural Priority Areas. The National Bank for Agriculture and Rural Development (NABARD) is providing the facility for automatic refinancing of the bank.

The program approves four models of Family Type Biogas Fertilizer Plants:[40]

- Pre-fabricated model biogas plants
- Floating Dome type biogas plants
- Bag type biogas plants (Flexi model)
- Fixed Dome biogas plants

(NBMMP), http://www.indg.in/rural-energy/schemes/national-biogas-and-manure-management-program-nbmmp (last visited 2 May 2012).

[39] Objectives of the National Biogas and Manure Management Programme (NBMMP), http://www.indg.in/rural-energy/schemes/national-biogas-and-manure-management-program-nbmmp (last visited 2 May 2012).

[40] Ibid.

The program applies to:[41]

- Cooking: Biogas can be used in a biogas chulhas/burner for cooking. A biogas plant of 2 m3 capacity is sufficient for providing cooking fuel to a family of four persons.
- Lighting: Biogas can be also used for lighting a biogas lamp for indoor or outdoor use. The requirement of gas for powering a 100 candle (60 W) lamp is 0.13 m3 per hour.
- Power Generation: Biogas can be used to operate a dual fuel or 100 per cent biogas engine and can replace up to 80 per cent of diesel in dual fuel engines.
- Refrigeration: Biogas can also be used for cooling applications in operating the chilling machines.

The program has the following monitoring mechanism.[42] A three-tier system exists to monitor the program which includes self-reporting by state governments and implementing agencies and involves 100 per cent physical verification of biogas plants at the block level. The second tier has random verification by nodal departments, implementing agencies and KVIC at the state level. The third tier has Regional Biogas Development and Training Centers of MNRE conducting inspections on a random basis.

The financial incentives being given by the Government are:[43]

- Central subsidy;
- Turn-key job fee linked with five years' free maintenance warranty;
- Financial support up to 50 per cent of the applicable Central Financial Assistance (CFA), subject to sharing of 50 per cent of the cost of repair by the beneficiary concerned for the repair of non-functional plants of more than five years old;
- Training of users, masons and entrepreneurs;
- Administrative charges to state government departments and agencies implementing the program;
- Regional level Biogas Training Centers; and
- Publicity, communication/extension.

---

[41] The National Biogas and Manure Management Programme (NBMMP), Ministry of New and Renewable Energy, http://www.mnre.gov.in/schemes/decentralized-systems/schems-2/ (last visited 2 May 2012).
[42] Ibid.
[43] 'Probable Questions for Biogas Programme,' Ministry of New and Renewable Energy, http://mnre.gov.in/file-manager/UserFiles/faq_biogas.htm (last visited 2 May 2012).

The cost of a biogas plant varies from place to place and with the size of the plant. The average cost of a 2 cubic meter size biogas plant is about Rs. 17,000/- varying with the height of the location – about 30 per cent more in hilly areas and about 50 per cent more in the North Eastern States. This affects the subsidies given by the Government. The central subsidy is given in fixed amounts for different categories of areas, states or regions. It varies from Rs. 4000/- to Rs. 8000/- per plant for general category states and Rs. 14,700/- per plant for the North Eastern States including Sikkim except the plain areas of Assam. Biogas Development Training Centers are providing technical and training support for the revival of non-functional plants nodal departments and receive 50 per cent subsidy from the Government to repair non-functioning plants.[44]

2. New Initiatives of the Program[45]

The Ministry launched another program on biogas-based distributed/grid power generation in January 2006 (2005–06) so as to set up reliable decentralized power generating units (3 kW to 250 kW) in rural areas in the country. The per kW CFA of Rs. 40,000 (3–20 kW), Rs. 35,000 (>20 to 100 kW) and Rs. 30,000 (>100 to 250 kW) is available for the installation of biogas-based power generation units. The program is implemented through nodal departments/agencies of the states through institutions and Biogas Development and Training Centers of the states' Union Territories and the Khadi and Village Industries Commission described above.

During the year 2008–09, the Ministry took up a new initiative to demonstrate an integrated technology-package in entrepreneurial mode on medium size (200–1000 m3/day) biogas fertilizer plants (BGFP) for the generation, purification/enrichment, bottling and piped distribution of biogas. Installation of such plants aims at meeting stationary and motive power, cooling, refrigeration and electricity needs in addition to cooking and heating requirements. Another request for proposals asking for an Expression of Interest for setting up such plants having capacity of above 1000 m³ was issued.

---

[44] 'Probable Questions for Biogas Programme,' Ministry of New and Renewable Energy, http://mnre.gov.in/file-manager/UserFiles/faq_biogas.htm (last visited 2 May 2012).

[45] The National Biogas and Manure Management Programme (NBMMP), Ministry of New and Renewable Energy, http://www.mnre.gov.in/schemes/decentralized-systems/schems-2/ (last visited 2 May 2012).

D.   USES

Biogas comes in useful as a cooking gas where women earlier used traditional methods of cooking in a clay oven called a 'chulha' where dried
cakes of cattle dung are burned to heat up the utensils. This led to the
creation of a lot of smoke, causing respiratory diseases to the women who
would cook on these ovens on a daily basis. Biogas also took away the
need for women to forage for fuel wood, which was a demanding task,
depriving them and their children the opportunity for work and education, thus reducing pressure on the forests. The other advantage of biogas
is that of sanitation. Toilets are linked to the biogas plant thus allowing for
a cleaner rural environment.

E.   PERSONNEL TRAINING, PUBLIC
PARTICIPATION PROVISIONS

1.   Promotion and Implementation of Biogas Technology in India[46]

The two rural-based renewable technologies, biogas and smokeless
biomass stoves, took India a few steps ahead in creating a sustainable
rural environment. Biogas as a technology had been disseminated under
the Khadi and Village Industries Commission (KVIC), using its floating
steel gas holder model in 1960. This took a step ahead when the Ministry
of Non-Conventional Sources of Energy (MNES) launched a central
scheme known as the National Project on Biogas Development (NPBD)
in 1981–82.

This scheme employed the Multi-Model and Multi-Agency approach,
thus allowing NGOs to be a part of the program. This constructive
engagement with NGOs has allowed India to install 3 million such biogas
plants through the country and in the process it has achieved its target of
having a nationwide presence of such a technology. Yet in the year 2000,
a new ambitious target was set up of installing 20 million plants. With the
implementation rate at 150,000 units per year, this figure will be achieved
in only 75–100 years.

The initiative also launched a few pilot projects for community size

---

[46]   Raymond Myles, 'Implementation of Household Biogas Plants by NGOs
in India,' Integrated Sustainable Energy (INSEDA) and Ecological Development
Association,   November   2001,   http://www.inseda.org/Additional%20material/
Lessons%20learnt%20NGOs%20Biogas%20program.pdf   (last   visited   2   May
2012).

biogas plants for villages which would provide fuel for domestic cooking and electrification for the village, along with generation of electricity to fulfill the needs of the local community. Successes have vastly eluded this scheme and if there have been any, they have been few and far between.

2. Experience of Implementation of Household Biogas Plants and Other Renewable Energy Technology by INSEDA'S NGO Network

The project is being implemented by the Integrated Sustainable Energy and Ecological Development Association (INSEDA) which is an official registered organization that is made up of more than 75 NGOs who have come together to promote renewable energy programs since 1980. INSEDA is headquartered in New Delhi with its members active in almost every state in India. Before INSEDA was formalized in 1980, it worked as an informal group for 15 years.

INSEDA's main goal is to develop and promote rural renewable sources of energy and create development in villages based on sustainability where the local community would actively participate to promote eco-friendly and environmentally sound sustainable human development. The deeper commitment of INSEDA is to create affordable forms of renewable sources of energy for the rural community, create ecological development which is environment friendly, along with the creation of suitable infrastructure in such communities which would support the first two endeavors.

The role played and the achievements in promotion and implementation of biogas and other renewable energy technologies by this NGO network for over two decades has been divided into five major stages, as summarized below:

a. First stage (1979 to 1981): recognition and promotion of household biogas plants by NGO groups

The recognition and potential of biogas by a group of Indian NGOs involved in agricultural and rural development in the early 1970s developed as follows:

i.   The NGOs came to a decision that biogas could be the viable fuel to answer the issues of domestic energy problems. Biogas could be integrated into the lives of the rural community seamlessly because the essential ingredients for the energy came from the dung (manure) of domesticated agricultural animals. This allowed women to have access to clean energy while providing a natural fertilizer for the farmer in the form of slurry, the by-product of a biogas plant.

ii. A census on bovine animals conducted in the 1970s stated that India
     had 240 million such animals while an average of 1000 million tons of
     dung was being produced. Of this, 300 million tons of dung was being
     used as dried cakes in traditional kitchen ovens for cooking purposes.
     The rest was left to decompose unscientifically, creating pollution
     and the release of methane (CH_4) and carbon dioxide (CO_2) into
     the atmosphere. During this time the biogas potential was said to be
     12–15 million tons based on the bovine population.

b.   Promotion and transfer of low-cost biogas plants by NGOs
The Janata model is an affordable fixed dome biogas plant. It was promoted by NGOs over the existing floating steel gas holder that was used
by the Khadi and Village Industries Commission. This new model was a
result of the research done by an Indian biogas research center. It used
brick and cement mortar, required local skills to be built, and was 30 per
cent cheaper than the KVIC model. The NGOs promoting this technology
then decided to train rural masons to construct the Janata model. They
also provided for transferring the skills and knowledge of the plant func-
tion to the owners. These NGOs also decided to form a network amongst
other NGOs to promote the cheaper biogas plants with the intention of
creating a national impact.

**c.   Second stage (1982 to 1984): popularization and extension of the
      Janata Biomass Model and development of a new low-cost fixed dome
      household biogas plant model by the NGO network**
The objective of this phase was to create stronger infrastructure at the
grassroots level in different parts of India. There was a need to create an
informal network of training programs which could disseminate the information and organize more workshops. At the same time the objective was
also to increase the number of NGOs in this field. The program also wanted
to launch the first phase, creating a network of financial support systems
aided by an overseas funding agency in mid-1983. Another objective was
to establish and operate Biogas Extension Centers (BECs) by each NGO
member of the network to put in place the plans and programs of the first
phase. Under the aegis of the National Project of Biogas Development of
the Ministry of Non-Conventional Energy (MNES), there was to be a systematic transfer and extension of household biogas plants in rural areas.
Planning and training were to be imparted to both the makers of the biogas
equipment and its users. More than 5,000 masons were trained to construct
fixed dome biogas plants along with training programs for users of biogas
plants. The final objective of the second phase was to have a participatory
development and evolution of a new low-cost fixed dome household biogas

plant of five different sizes (1, 2, 3, 4 and 6 m3 capacities) by the NGO network, which was christened as the Deenbandhu (meaning, friend of the poor).

d.   Third stage (1985 to 1989): field evaluation, demonstration, promotion, transfer, popularization, dissemination, training and extension of the Deenbandhu model by the NGO network

This stage saw the evaluation of the different models of the Deenbandhu Biogas Plant (DBP) after on-the-field analysis of the various models which helped in refining the final design. The MNES approved the Deenbandhu plants in 1986–87 and they were put under the National Project on Biogas Development (NPBD). From 1983–89, the dissemination of DBP was done by NGOs using funding by an overseas agency. In the first phase 42,000 household biogas plants were made which were funded by Indian resources. Of the costs of these plants, 30 to 50 per cent were met by subsidies provided by NPBD (MNES) while the rest was taken up by the owner or by taking loans. State governments also provided some subsidies. The ratio of Indian funding as compared to overseas investment in building these biogas plants came to 1:5. Experiments were also conducted with the present biogas plants using alternative building materials by the network of NGOs, while the current technology was also transferred to other developing countries.

e.   Fourth stage (1990 to 1995): implementation of phase two of the biogas program by the NGO network

The second phase of the biogas dissemination program constructed 35,000 household plants (mainly the DBP model) and gave maintenance services to plants already owned. The Government also created capacity building measures in the existing NGO network while adding more members. Research and development was pursued to make existing biogas plants better and to design newer low-cost biogas plant models. Additionally research was conducted on the utilization of slurry to increase crop production and it was concluded that it was better than inorganic fertilizer, along with saving the farmer the considerable sum of money which went into its purchase. The Deenbandhu model biogas plant was exported to other developing countries. The program facilitated decentralization of the network by promoting Regional Consultative Groups for solving regional level problems.

The biogas network of NGOs was decentralized on a regional basis, giving autonomy to these regional bodies. In 1992 there were 60 members, operating 75 Biogas Extension Centers (BECs) and by 1993 the number rose to 70 members, operating 90 BECs. One such regional group was called Sustainable Development Agency and received direct funds from MNES under NBPD. By 1992 these autonomous bodies were formalized into a registered body called Integrated Sustainable Energy and Ecological Development Association (INSEDA) which also pursued renewable energy technologies. They sought to keep up with the ever developing field of renewable sources of energy and promote them in a systematic way. Another innovation researched on biogas plants was to use bamboo instead of bricks as the key building agent for the fixed dome variety. By 1995 a total of 85,000 biogas plants had been built, 42,000 during Phase 1 and 43,000 during Phase 2.

f.   Fifth stage (October 1996 to 2001)
By 1996, the 50 members of INSEDA had built 100,000 household biogas plants. This number rose in 2001 where 75 such NGO members built 130,000 household plants under the NPBD. The INSEDA was involved in promoting renewable sources of energy by providing technical expertise and being a facilitator of the same. The focus was on low-cost household biogas plants. In 2001 INSEDA, on the suggestion of an NGO, built 1000 Deenbandhu model low-cost biogas plants using ferro-cement technology instead of brick and mortar. The low-cost biogas plant made of bamboo reinforced cement mortar was christened as Shramik Bandhu (meaning, friend of the labor). It was a labor intensive design with labor constituting 45 per cent of the cost. The final design was approved by 1996 and this model was targeted towards rural women, landless peasants, artisans, masons, unemployed rural youth and other poor villagers who earn their living through working as daily wage laborers in rural areas. Its use by them required some training.

INSEDA in collaboration with the Foundation for Alternate Energy (Slovakia) (FAE), and with financial support from the International Network for Sustainable Energy (INFORSE), prepared material for Distant Internet Education on Renewable Energy Technology (in short known as DIERET) and launched an international program on DIERET materials through the Internet. The pilot phase of implementation of DIERET was launched jointly by FAE and INSEDA through the Internet in April/May 2000, which had a good response. The first group of trainees completed their programs by August 2000. INSEDA and FAE are now looking for funding agencies and sponsors to launch DIERET for the benefit of NGOs in developing countries on a regular basis.

F. LESSONS LEARNED FROM THE IMPLEMENTATION OF LOW-COST DECENTRALIZED RENEWABLE ENERGY PROGRAMS IN RURAL AREAS OF INDIA

Renewable energy technologies must be understood by the rural population before they are implemented and transferred. This means that such technology must be demonstrated and promoted adequately before putting it into action. As failure of any such innovation has a negative impact on the whole program in the immediate rural community and its vicinity, the application of such technology has to be vetted first before it is put into action. Established local NGOs must aid in such demonstrations. New technology must be affordable and this calls for a greater gestation period which allows for acceptance and internalization by the local people in rural areas. Renewable energy technologies (RETs) further must be used as a means to empower the local community. With this end goal they can create a 'sustainable RET market'. RETs can also create more jobs by training the youth, allowing them to build capacity in the village and making the village a self-functioning unit which is managed by the people of the community. RETs also must be amalgamated with other development programs which would lead to 'sustainable human development', which when linked with quantitative and qualitative growth empowers people and could become a self-sustaining marketable object.

G. IMPORTANT CONSIDERATIONS FOR THE SUCCESS OF PEOPLE-ORIENTED RENEWABLE ENERGY-BASED RURAL DEVELOPMENT PROGRAMS

With India's massive social, economic, linguistic, topological and cultural diversity it becomes important to recognize that RETs should cater to the communities individually, respecting their traditions. Before promoting RETs and RET-based development programs the Government or other sponsors need to engage in a process-oriented approach, comprising education, training, demonstration, awareness, capacity building, technical literacy and skills development in renewable energy. This is a trust building exercise allowing the program to forge a partnership with the local community which gives the renewable source of energy a better opportunity to be accepted, adopted, assimilated, absorbed and internalized as a sustainable energy option for a better future of the local people and communities. The local community must be treated as a stakeholder of the project and not merely as a beneficiary. RET programs should also create employment and opportunities for self-employment. The local community must be taught about the technology operating the plants so that they can operate, manage, maintain, service and repair equipment locally, and spare parts should be easily available. There is also a need to integrate energy programs with other development programs of the village and other non-economic pursuits which will help promote sustainable human development.

A worthy pursuit could be to integrate activities involving renewable sources of energy with that of eco-food production to regenerate the micro-agro-eco system. NGOs must establish RET Resource Centers linking RETs with entrepreneurship development and providing sustenance to marginalized populations by providing them with education and training in repair, management and development skills. This will go a long way in providing empowerment to people who need it, like women, rural unemployed youths, landless peasants, local artisans and masons.

H.   CHALLENGES AND OPPORTUNITY

Building a biogas plant which is too large could create issues if the household has limited use for the gas the plant produces. This could happen either by accident or on purpose. Households have a general tendency to install oversized plants while they have limited use for cooking purposes. These households do not apply the plant for other extensive energy requirements. These large biogas plants are constantly underfed which leads to lower production of gas. These plants are also underfed because of undercollection of dung ranging to about 30–40 per cent of the required capacity. Undercollection of dung also accounts for issues with biogas plants. This happens when cattle are made to work in the fields. Climate changes and areas prone to droughts cause cattle to die or be sold off. This produces less cattle dung which also leads to undercollection of dung. In some areas, the plant may not be technically feasible all year round due to low winter temperatures that inhibit methanogenesis.[47]

Faulty construction of biogas plants also poses a challenge especially when the plant is constructed by workers who are not experts in the craft in order to reduce capital costs. Fixed dome biogas plants which come

---

[47]   Jo Lawbuary, 'Biogas in India: More than Gandhi's Dream?,' http://www.ganesha.co.uk/Articles/Biogas%20Technology%20in%20India.htm (last visited 2 May 2012).

pre-cast often develop faults due to their construction. These problems later compound when the operating plant becomes non-functioning due to shoddy construction work. On many occasions, the services of skilled masons who are trained to install biogas are discarded due to the higher charges demanded by them. In their place trainees with limited technical knowledge and more often than not little practical skill are forced to meet ambitious targets set by local and governmental agencies; a practice which is heavily encouraged. In a study done on such non-functional plants in the western state of Maharashtra, surveyors found that of the 1670 studied, 50 per cent of them were assessed to be incapable of ever being made functional. Additionally the targets set by the Government have always been too ambitious and coupled with the lack of resources, and lack of local decision-making, hurts the potential of achieving targets. It should also be recognized that biogas is not an effective alternative for many rural families because the requirements to install a biogas plant are many and substantial. Apart from the number of cattle required, each family requires 50 kilograms of dung every day along with 50 liters of water, a permanent home and the ability to make the initial investment. These are daunting targets for a family which is not well-to-do. Studies have shown that community biogas plants with public utility as their end are far more economical than household biogas plants, leading to speculation that biogas plants could be linked to small-scale industries.[48]

The larger economy of subsidized conventional energy such as electricity, along with free connection to the grid for farmers, will keep these forms of energy at a lower cost. Bringing parity of cost with conventional sources of energy requires far more subsidies for fuels like biogas or warrants a rise in price for conventional fuels. The system of grants and loans may hinder the correct choice of plant for different users, such as the ineligibility of community size systems, due to their size. While finally, another point in prohibiting uptake may be the perceived unnecessary switch from the existing free source of energy, such as wood and crop residues.[49]

Cultural practices also slow down the use of biogas, like the fact that the traditional bread of Indian households cannot be roasted well in the burners fueled by biogas or that basic lentils take a longer time to cook. These are the staple diet for every Indian family, barring cultural and economic divisions. Another hindering cultural practice could be that of decision-making wherein the male members of the family might have other priorities than those of the household and the woman. It has also been

---

[48]   Ibid.
[49]   Ibid.

established that there is deforestation from the collection of firewood as an alternate source of fuel, and so the Government must brand biofuels differently to make them appealing to households.[50]

Criticism of national policies has been wide such as the lax selection process, and the arbitrary fixing of regional targets, which are then pursued. The problems mentioned above could be dealt with effectively if the Government of India can institutionalize the selection process of technology, extension and support services. The Government of India, though, has set ambitious targets and will not be able to attain them for lack of organization. The National Project on Biogas Development (NPBD) has been widely criticized for the unscientific ways of fixing regional targets and the selection process, which has been negligent. In a study of biogas plants in the state of Maharashtra, it was found that a majority of the holders of such plants were not supposed to be qualifying under the feasibility criteria. Though biogas plants have reached many corners of the country, but without adequate technical support, functioning plants will soon run their course and stop for good. If this happens, biogas in India will be the elitist dream and not an initiative of educated NGOs and policy makers in an otherwise rural and uneducated setting. These can be countered through effective selection processes for the technology, and proper extension and support services. Lack of technical ability has been a recurring issue with dysfunctional plants. Every village must be given adequate maintenance tools. There is a need to assess the distinct needs of each rural community. Owing to the huge amount of diversity in India, a singular plan will not suit every need. The participation of people of the community must be emphasized. This will ensure their interest in the upkeep of the biogas plants. The organizations must also focus on educating the local community about the biogas technology, and committing them towards planning and monitoring of the plants thus making the technology viable and sustainable for the community at large. Co-ordinated management information systems could also be another viable option in this scenario. Comparisons with the Chinese program have always existed and knowing their success story one must consider those inputs for India. Emphasis on training and education has allowed biogas to be disseminated widely. A strong social organization may particularly facilitate the spread of new, community-focused technologies. Focus must be on micro-planning and on grassroot issues and realities. Community driven participation in rural India is the key to success of this program. Renewable sources of energy like biogas can attain success only

---

[50] See Gustavsson, supra note 22.

if the people are educated about its benefits. Co-ordinated management information systems must be made part of biogas development, in order for problems to be identified and remedial measures undertaken.[51]

There is lack of data when it comes to biogas. There is difficulty in quantifying success in the area. Further one needs to not only quantify functioning and dysfunctional biogas plants to gauge the success of the program but also look at quantity and quality of the gas and effluent produced by the plant. There is also a need to quantify the benefits of the different types of biogas plants. The other issue which sidetracks the use of biogas is the current subsidies given to conventional forms of energy in India. Thus the 'macro environment' which determines the price structure of conventional fuels creates a disincentive for people to use unconventional fuels. Unless subsidies being given to biogas are not brought in line with those of conventional fuels along with free grid connection to farmers cut off, biogas will not remain the preferred source of energy in rural India. Another hindrance is the reluctance to leave free sources of energy like fuel wood, dung cakes and crop residue. The Indian system of grants and loans also causes different types of biogas to be outside their ambit and thus creating issues with possible owners who in the first place were willing to install them.[52]

I. CONCLUSION

Biogas remains a marginal fuel even though it shows tremendous potential in rural India, which is due to the logistical difficulties associated with it along with the fact that the fuel is disseminated without proper channels and there are many unattended dysfunctional plants strewn across the landscape of the country. Biogas, though useful to create a prudent rural economy, has found its application to be logistically difficult. Ambitious targets set and a loosely held organization to support its dissemination have resulted in very high rates of non-functioning plants. As such there is a reluctance shown by members of rural communities to employ this source of energy thus making its impact marginal at best. Biogas must be made user-friendly if it has to inspire villages across rural India to take to the biogas. Though it promises a lot, its advantages to the marginalized have been little. With India's socio-economic diversity, the biogas

---

[51] Ibid. See also David Wargert, 'Biogas in Developing Rural Areas,' Lund University, 2009, http://www.davidwargert.net/docs/Biogas.pdf.

[52] See Gustavsson, supra note 22.

program was looking to help the poorest class of India, the 'scheduled caste' and 'scheduled tribe', yet due to the already marginal status of these classes, the program has not been able to help them tangibly as was expected. At the same time these distressed classes have to forage for, the once free, dung which has now fallen prey to heavy 'commodification' since biogas was introduced to rural communities. With the commodification of dung, the poorest of families find it difficult to vouch for the good that comes from a biogas plant. What was once free now comes with a price tag and as such this robs biogas of the label of being a sustainable rural fuel. With demand for sustainable fuel peaking in rural areas, biogas has become an anonymous research subject, having not been taken up by major journals after the 1990's to research on. Though dissemination of biogas is still being undertaken, its situation is not completely known.[53]

[53]   See Lawbuary, supra note 18.

6. Case study of renewable energy in Brazil

Richard L. Ottinger with Douglas S. de Figueiredo and Lia Helena M.L. Demange

A. INTRODUCTION

1. General Overview of the Country's Energy Situation

Brazil is the largest (8,514,215.3 sq km, 3,287,357.26 sq miles)[1] and most populous country (190,732,694)[2] of South America and is facing strong economic growth. It is now the largest economy of the region, and it is expanding its presence in world markets.[3] The average expected growth of the Brazilian Gross Domestic Product (GDP) for the next 6 years is 5 percent[4] and a reliable energy sector is extremely important to maintain that growth path.

Brazil has been using its natural resources to generate renewable energy, mostly with hydropower to generate electricity and its unique sugar cane fuel program (Proálcool) to run vehicles. The high utilization of hydropower limited the contribution of other sources for the electrical base load; thus Brazil is below the world average in use of nuclear, natural gas and coal for electricity production. But with the recent discoveries of large

---

[1]   Instituto Brasileiro de Geografia e Estatistica, Censo 2000, available at http://www.ibge.gov.br/home/estatistica/populacao/default_censo_2000.shtm (last accessed 4 April 2013).

[2]   Instituto Brasileiro de Geografia e Estatistica, Censo 2010, available at http://www.ibge.gov.br/home/estatistica/populacao/censo2010/default.shtm (last accessed 4 April 2013).

[3]   Euromonitor International, http://www.euromonitor.com/pdf/2020rank. pdf (last accessed 12 November 2010).

[4]   Ministério da Fazenda [Brazilian Treasury Department], Economia Brasileira em Perspectiva, available at http://www.fazenda.gov.br/portugues/docs/ perspectiva-economia-brasileira/edicoes/Economia-Brasileira-Em-Perspectiva-Jan10.pdf (last accessed 12 November 2010).

offshore oil deposits,[5] Brazil is likely to become one of the largest oil producers in the world, which might cause major changes to the national patterns of electricity generation.

In the 1990s, the Brazilian government privatized most of its electricity industries. Before privatization, energy planning was mandatory. After privatization, energy planning became merely recommended for the energy industry, not binding. To a large extent, the industry did not follow the targets of generation growth suggested in the government's decennial plans.

In the late 1990s, Brazil faced a severe drought that led to a serious depletion in reservoir levels. As the country did not carry out diligent energy planning in the previous years, it was not able to generate sufficient power to meet its demand in 2001. The result was a drastic government-ordered cutback in usage which included the imposition of usage quotas. Nevertheless the country experienced rolling blackouts.[6]

The vulnerability of the energy sector to the late 1990s drought shows the threat to security of the electricity supply the Brazilian government faces due to the high dependence of its energy on seasonal conditions.

After the rolling blackouts, the federal government decided to encourage the diversification of energy sources that included efforts to increase electricity generation by utilization of natural gas, nuclear, biomass, wind, solar (photovoltaic and thermal) and small hydropower resources. As a result, between 1998 and 2008, Brazil installed generating capacity that more than doubled, from 49.6 GW to 102.61 GW,[7] and the participation of hydropower in the country's generation of electricity changed from 95 percent in 1997 to 76.9 percent in 2009.[8]

Due to the 2000/01 blackouts, the Brazilian government created the Brazilian Energy Research Company,[9] a legal entity under the Ministry of Mines and Energy, with the objective to provide study and research to support energy sector planning.

The Brazilian Energy Research Company is composed of representatives of all energy-related sectors, including alternative energy generators and oil producers. In 2007, this entity issued, in collaboration with the Ministry of Mines and Energy, the Brazilian National Energy Plan

[5] Presalt, http://www.presalt.com (last accessed 12 November 2010).
[6] 2010 Brazil Energy Handbook, PSI Media Inc., December 2009.
[7] 2010 Brazil Energy Handbook, PSI Media Inc., December 2009.
[8] 'Balanço Energético Nacional,' 2010, Ministério de Minas e Energia, https://ben.epe.gov.br/BENRelatorioFinal2010.aspx (last accessed 13 April 2013).
[9] Bylaw 10.847 of 15 March 2004, regulated by Decree 5.184 of 15 September 2004.

(BNEP), which was the first comprehensive plan of energy resources made by the Brazilian government; it embraces the long-term national energy framework.

This plan identifies the requirements and tendencies of the energy sector, shapes the alternatives of expansion of the sector, and indicates the problems that the country might face in the next couple of decades; it also is a forum for new discussions related to challenges and opportunities in relation to the energy sector, establishing benchmarks for each segment of the sector.

According to this plan, hydroelectric generation is planned to continue its expansion with the construction of two gigantic plants on the Rio Madeira in Rondonia (Santo Antonio and Jirau facilities, 3.2 GW and 3.3 GW respectively).

These two hydro power plants are expected to help Brazil to meet its electricity demands in the mid-term. For the long term, the balance of hydroelectricity demand should be met in part by the 11.2-GW Belo Monte dam.

Brazil is also planning to increase the use of wind technology. In December 2009, Brazil held its first supply tender exclusively for wind farms. In the event, 1.8 GW of capacity were purchased for development by mid-2012. However, wind remains a modest component of Brazil's renewable energy mix sources, responsible for just 0.2 percent of national electrical generation in 2009.[10]

Brazil also plans to expand its sugar cane production, estimated to increase from 420 to 1,140 million tons per year. Ethanol production based on the BNEP is predicted to increase at a 5 percent per year rate that will dominate the gasoline/ethanol demand for vehicle transportation. However, the ethanol contribution to electricity generation is still small, representing 5.4 percent of national electrical generation in 2009.[11]

The next BNEP is expected to include the use of firewood and charcoal for residential purposes and by the metallurgical industry. The use of fire-wood and charcoal for meeting residential energy needs, such as cooking and heating, causes the death of approximately 2.5 million women and children every year around the world due to inhalation of smoke.[12] In

---

[10] 'Balanço Energético Nacional', 2010, Ministério de Minas e Energia, https://ben.epe.gov.br/BENRelatorioFinal2010.aspx (last accessed 13 April 2013).
[11] Ibid.
[12] UNDP, UNDESA and World Energy Council (2002) quoted in ITDG and Greenpeace (2002), 'Sustainable Energy for Poverty Reduction: An Action Plan, Greenpeace and the Intermediate Technology Development Group', 3, Greenpeace, http://www.greenpeace.org/international/Global/international/plan et-2/report/2006/3/sustainable-energy-for-poverty.pdf (last accessed 13 April 2013).

urban areas, poor people often use wood waste from civil construction for cooking and heating due to the scarcity of wood. This situation is a serious threat to women's and children's health caused by the toxic smoke resulting from the fire and particularly from the combustion of paint and varnish present in the wood.[13]

The use of firewood and charcoal, and also kerosene for lighting, also increases poverty, because people, mainly women and children, have to spend a lot of time collecting fuel every day,[14] thereby losing the opportunity for education and income-producing activities for themselves and their communities.[15] Considering the time spent on the collection or the money disbursed on the purchase of wood, charcoal and kerosene, it is estimated that up to a third of poor people's income is spent on energy.[16]

The use of wood as a source of energy also encourages illegal deforestation that seldom is followed by reforestation; therefore, serious degradation of forest ecosystems occurs, and this is a substantial source of greenhouse gasses that are the cause of global climate change.[17]

B.   NEW RENEWABLES IN THE ELECTRICITY MARKET

The Brazilian private electricity market was created after the privatization of the energy sector in the mid-1990s. The market is divided into the regulated market and the free market. The regulated market is composed of the total electricity necessary to supply the captive market, which is the

---

[13]   José Goldemberg and Oswaldo Lucon, 2008, 'Energia, Meio Ambiente e Desenvolvimento' [Energy, Environment and Development], São Paulo: Edusp, p.60.

[14]   In India, two to seven hours each day can be devoted to the collection of fuel for cooking. UNDP, UNDESA and World Energy Council (2002) quoted in the International Energy Agency's World Energy Outlook 2002, 366, World Energy Outlook,   http://www.worldenergyoutlook.org/media/weowebsite/2008-1994/weo2002_part2.pdf (last accessed 13 April 2013).

[15]   Roland Bunzenthal, April 2007, 'Battling against Poverty' in Financing Development, KFW Entwicklungsbank in cooperation with Germany's development bank, pp.II–III.

[16]   ITDG and Greenpeace, 2002, 'Sustainable Energy for Poverty Reduction: An Action Plan, Greenpeace and the Intermediate Technology Development Group', 7, Greenpeace, http://www.greenpeace.org/international/Global/international/planet-2/report/2006/3/sustainable-energy-for-poverty.pdf (last accessed 13 April 2013).

[17]   Édis Milaré, 2009, 'Direito do Ambiente' [Environmental Law], São Paulo: RT, p. 1302.

market formed by the consumers that buy energy from energy distribution facilities. Large consumers of energy have the option to purchase energy in the free market directly from the generators, without the intermediation of energy distributors. The most important difference between these two market environments is that the regulated market involves stricter governmental oversight of the allocation of energy supply; and it obtains its electricity through tenders, while the free market acquires its electricity through negotiation.

There are different sorts of incentives for the commercialization of renewable sources in each market. In the regulated market, the energy distribution facilities are obliged by law to buy the energy generated within a Program of Incentives for Alternative Electricity Sources, PROINFA, which will be explained in detail below. The regulated market also promotes tenders exclusively offering new renewable energy for the acquisition of electricity needs to ensure the security of supply. By these means, the Brazilian government avoids direct competition between new renewable energy and conventional energy in the market. PROINFA was expected to add nearly 3,300 MW to the national electrical grid through the end of 2011. The tenders for modern renewable sources amounted to 8,837.3 MW between 2007 and 2011, from which 541.9 MW already entered the grid in 2010; 2,379.4 MW were expected to enter the grid by the end of 2012; the rest is to be supplied no later than 2014.[18]

---

[18]   EPE, Empresa de Pesquisa Energética, '1° Leilão de Energia de Fontes Alternativas agrega 638.64 MW ao SIN' [1st Auction of Modern Renewable Energy adds 638,64 MW to the National Interconnected System], Rio de Janeiro: 18 June 2007, http://www.epe.gov.br/leiloes/Paginas/Leil%C3%A3o%20de%20 Fontes%20Alternativas%202007/LeilaoFA2007_33.aspx?CategoriaID=43 (last accessed 13 April 2013); EPE, Empresa de Pesquisa Energética, 'Leilão de Energia de Reserva negocia 2.379 MW de térmicas à biomassa' [Auction of Storage Energy comercializes 2,379 MW from thermo plants fueled by biomass], São Paulo: 14 August 2008, http://www.epe.gov.br/imprensa/PressReleases/20080814_1.pdf (last accessed 13 April 2013); EPE, Empresa de Pesquisa Energética, 'Primeiro leilão de energia eólica do país viabiliza a construção de 1.805,7 MW' [First national auction of wind energy makes the installation of 1,805.7 MW possible], São Paulo: 14 December 2009, http://www.epe.gov.br/imprensa/PressReleases/20091214_1. pdf (last accessed 13 April 2013); EPE, Empresa de Pesquisa Energética, 'Leilões de Fontes Alternativas contratam 89 usinas, com 2.892,2 MW' [Auctions of Modern Renewables contracted 89 power plants, which totalize 2,892.2 MW], São Paulo: 26 August 2010, http://www.epe.gov.br/imprensa/PressReleases/20100826_1.pdf (last accessed 13 April 2013); EPE, Empresa de Pesquisa Energética, 'Contratação no Leilão de Reserva totaliza 1.218,1 M.W, através de 41 usinas' [Contracts celebrated during Auction for Storage Energy totalize 1,218.1 MW, generated by 41 power plants], São Paulo: 18 August 2011, http://www.epe.gov.br/imprensa/ PressReleases/20110818_1.pdf (last accessed 13 April 2013).

In the free market, the purchase of energy from new renewable sources is stimulated by tax deductions for the use of transmission lines.[19] Also, free market consumers of alternative sources of energy are allowed greater flexibility to choose whether to participate in the regulated or in the free market[20] that gives them greater capability to respond to changes in energy market prices. Finally, the minimum consumption rate for leaving the captive market and entering the free market is lower for consumers of renewable energy.[21] That means that medium size consumers, such as shopping malls and other commercial facilities, can only participate in the free market if they purchase renewable energy from decentralized energy generation sources, the so called 'alternative sources.' The Brazilian law classifies as alternative sources wind, biomass and small hydropower plants.[22]

Parallel to these incentives, modern renewables receive other sorts of tax reductions that are applied to energy in both markets: wind and solar energy are exempt from taxation on the circulation of goods;[23] and wind generators are exempt from taxation on imported goods.[24] The federal government estimated a reduction of collection of taxes in R$ 89 million (US$ 48.37 million) in 2010 due to the exemption of taxation for wind generators.[25]

C.   THE PROGRAM OF INCENTIVES FOR ALTERNATIVE ELECTRICITY SOURCES

1.   Brief Project Description

PROINFA was established as a government program in April 2002, revised in November 2003,[26] and its main objective is to diversify Brazilian energy generation resources and look for regional solutions to incentivize

---

[19]   Law n 9427/1996, article 26, §1 and Resolução Normativa Aneel n 77/2004, article 2.
[20]   Resolução Normativa Aneel n 247/2006, article 8, §1 and Law n 9074/1995, article 15, §8h.
[21]   Resolução Normativa Aneel n 247/2006, article 1, §1 and Law n 9074/1995, article 16.
[22]   Law n 10.438/2002, article 3, caput.
[23]   Convênio ICMS 101/97.
[24]   Decreto 7032/2009.
[25]   Ministério da Fazenda, http://www.fazenda.gov.br/portugues/releases/2010/janeiro/TABELA-MEDIDAS-29-01-10.pdf, (last accessed 4 March 2013).
[26]   Law n 10.438 dated 26 April 2002, revised by Law n 10.762 dated 11 November 2003.

the use of renewable energy sources, considering the economic aspect of the available inputs and technologies that might be applicable to increase the use of these renewable energy sources on the Brazilian grid.

PROINFA was created in two phases. The First Phase was launched to promote the generation of 3,300 MW from wind, biomass and small hydropower in equal shares. In this phase, the federal government subsidized the deployment of these energy sources by assuring that the energy produced in a 20-year term will be purchased by Eletrobrás, the government-owned electricity company, if the project is in accordance with the government guidelines and with its establishing law.[27]

In terms of incentives and benefits, PROINFA assures the project developers execution of 20-year long-term Power Purchasing Agreements (PPAs) with Electrobrás, which is in charge of the execution of the PPAs.

The tariff to be paid to the project developers must be calculated by the Ministry of Mines and Energy to assure the feasibility of the projects pursuant to a cash flow analysis that takes into account the following factors:

(a) 20 year long term commitment period; (b) rate of return compatible with the risks of the program; (c) efficiency standards, technological level and generation potential; (d) standard costs for each source; (e) estimated value of the facility at the end of the operation period; (f) estimates of operational costs, transmission fees and taxes; (g) estimation of unavailability periods and self-consumption of electricity; (h) financial conditions of the program; (i) equity-debt ratio compatible with the standards of the electricity generation market; (j) specific discounts on transmission and distribution fees; (k) depreciation rates applicable; and (l) income from sale of byproducts.[28]

The prices are subjected to a minimum price based on the average national energy charges of the last 12 months and are established according to the value of the technology, in which wind is determined to be 20 percent more expensive than small hydropower, which is determined to be 20 percent more expensive than biomass.[29]

The prices of energy generated by these three sources are updated annually with readjustment rates that are uniformly applied to all sources. Thus, the price of wind, small hydropower and biomass increased by 3.49 percent in 2006, 5.49 percent in 2007 and 12.24 percent in 2008.

The First Phase of the program prioritized the purchase of energy from 'autonomous independent producers' (PIAs), small energy companies that do not hold an authorization/grant/concession/right to develop a

---

27 Ibid.
28 Decree 5.025 of 30 March 2004, article 3.
29 Law n 10,438/2002 article 3(b).

public utility/services/franchise.[30] The energy produced by companies that do not meet such a classification can be purchased only if there is no PIA competing for the contract and only if the agreement does not comprehend more than 25 percent of the total energy acquisitions under the program.[31]

One of the main advantages of PROINFA is to ensure the economic and financial feasibility of electricity generation from renewable sources.[32]

PROINFA was a pioneer program that allowed Brazil in three years to increase its wind farm installed capacity from 22 MW to 414 MW. PROINFA also reflects the Brazilian commitment related to the international issue of climate change, playing a major role in the region and being an example to be followed.

In addition, the program was responsible for making energy generation by PIAs possible. Before PROINFA was launched, PIAs faced great problems in supplying energy because they could not compete with large energy companies, since such companies were entitled to exclusive benefits, such as reduced cost of energy transport.[33]

After PROINFA started, beside the benefits that PIAs received by having a guaranteed demand for the energy generated, they also became eligible for special investment programs of the National Bank of Economic and Social Development. The bank provides loans for up to 70 percent of all investments in modern renewable sources, except for expansions of acquisition of land and of imported services and goods. The bank also provided wind projects with longer deadlines for loan payment and smaller fees than other renewable sources, given that wind power projects present a lower return on investments than other renewables. These

[30] Law n 10,438/2002 article 3, §1.
[31] Law n 10,438/2002 article 3, §2.
[32] D. de Avila Vio, 'Oil, Gas & Energy Law Intelligence: An overview of Proinfa – A Brazilian Incentive Programme for Alternative Electricity Sources', http://www.ogel.org/article.asp?key=3079 (last accessed 13 April 2013).
[33] See Vilson Daniel Christofari, 'Fontes Alternativas de Energia: aspectos ambientais e estratégicos – segurança dos sistemas' [Modern Renewable Sources: strategic and environmental aspects – the security of the supplying systems], Revista do Direito da Energia, ano I, n 1, p. 184–197, abr. 2004; José Goldemberg and Oswaldo Lucon, 2008, 'Energia, meio ambiente e desenvolvimento' [Energy, environment and development], 3a ed, São Paulo: Edusp; Edmundo Emerson Medeiros, 'Infra-estrutura energética e desenvolvimento: Estado, Planejamento e Regulação do Setor Elétrico Brasileiro' [Energy Infrastructure and Development: State, Planning and Regulation of the Brazilian Electrical Sector], 2008, Dissertação de (Mestrado em Direito), Faculdade de Direito da Universidade de São Paulo, São Paulo.

measures have been successful in enhancing the wind power industry, as the offers to purchase wind energy in PROINFA's auctions have been greater than the offers and purchases of energy from other sources.

PROINFA also promotes the national renewable industry by establishing that the project to be developed and eligible under PROINFA must have at least 60 percent of Brazilian machinery and materials.[34] This condition represented an initial obstacle to the launch of wind power because the country did not contain national technology in this field. However, such disadvantage was compensated by the facilitation of loans and tax deductions to wind facilities.

The legality of governmental subsidies to boost national renewable energy industry is currently under dispute at the World Trade Organization due to a complaint filed on September 13, 2010 by Japan against measures taken by the Canadian province of Ontario.[35] A decision in favor of Japan would put Brazil in a vulnerable situation, as the country also adopts protective policies to subsidize its national renewable energy industry. The Brazilian government is aware of this situation, requesting the right to be a third party in the Japanese complaint.[36]

PROINFA was responsible for the development of 144 renewable energy power plants which represent installed capacity of 3,299.40 MW, with 1,191.24 MW occurring from 63 small hydro power plants, 1,422.92 MW occurring from wind farms and 685.24 MW occurring from 27 thermal power plants fueled by biomass. According to the law that established PROINFA, the energy generation facilities should have started providing energy by December 30, 2006.[37] This deadline was repeatedly extended, however, first to December 30, 2008,[38] then to December 30, 2010[39] and recently to December 30, 2011.[40]

Some of the causes in the delay of the operations can be attributed to the way that PROINFA was structured. Through this program, the federal government tried to solve three problems at the same time: stimulate the use of alternative energy sources; incentivize the development of the national industry; and introduce new players in the energy market.

---

[34]   Law n 10,438/2002 article 3, f, I.
[35]   Dispute DS412, World Trade Organization, Canada, 'Certain Measures Affecting the Renewable Energy Generation Sector', http://www.wto.org/english/ tratop e/cases e/ds412 e him (last accessed November 2011).
[36]   Ibid.
[37]   Law n 10,438/2002 article 3, I, a.
[38]   Law n 11,075/2004, article 4.
[39]   Law n 11,943/2009, article 21.
[40]   Law n 12.431/2011, article 21.

Because of this triple objective, the expansion of the use of alternative sources failed to be the principal priority and therefore implementation was delayed. The incentive to develop the national industry, in accordance with the requirements of article 3, I, f, of Law n 10.438/2002,[41] supported the delay in the installation of the renewable energy business, especially wind sources, because the national industry did not have enough supplies of machinery; that generated delays in delivery and price increases.[42]

The objective of including new lower economic players instead of established energy companies, as well as the lack of criteria by the government in the selection of the contractors and projects, resulted in the recruitment of players with a lack of technical and financial capacity. These factors contributed to the occurrence of problems and delays in the execution of the projects.

When the program was launched in 2002, it was predicted that after the goal of 3,300 MW was achieved, PROINFA would enter its Second Phase, with a goal to ensure that 10 percent of national energy would be generated from biomass, wind and small hydropower. However, there is some uncertainty whether the Second Phase will actually take place due to the delay in the execution of the First Phase and the silence of the government about what is going to happen to the program after the First Phase is finished.

D. WHY INITIATED

As a result of the Brazilian lack of an energy plan to address its growing demands, its dependence on hydroelectric power plants and the severe drought in 1990, Brazil faced a drastic electricity crisis with severe power shortages, and the government decided to impose a mandatory curb on consumption (which led to a significant reduction of GDP growth) and to

---

[41]   Article 3, I, f: 'Será admitida a participação direta de fabricantes de equipamentos de geração, sua controlada, coligada ou controladora na constituição do Produtor Independente Autônomo, desde que o índice de nacionalização dos equipamentos e serviços seja, na primeira etapa, de, no mínimo sessenta por cento em valor e, na segunda etapa, de, no mínimo, noventa por cento em valor.' ['The direct participation of manufacturers of generation equipment and its controlled or colligated companies is admitted in the constitution of the Autonomous Independent Producer, as long as the rate of nationalization of equipment and services in the first phase of the Program is of, at least, sixty percent in terms of value, and of, at least, ninety percent, in the second phase of the Program.']
[42]   Elisângela Mendonça, 'Adeus, Proinfa 2' [Goodbye, Proinfa 2], 2 fev. 2010, Brasil Energia, http://www.energiahoje.com/brasilenergia/noticiario/2010/02/02/403330/adeus-proinfa-2.html, (last accessed 9 August 2011).

incentivize the use of renewable energy to be displaced to the grid, launching PROINFA.

1. Public/Private Funding and Participation by Funders and/or International Organizations

The First Phase subsidies/incentives for the program were funded by a government Energy Development Account and by the sale of carbon credits generated for the program under the Clean Development Mechanism (CDM) of the Kyoto Protocol. PROINFA is predicted to reduce annually 3 million tons of carbon dioxide.[43] Consumers pay into the Energy Development Account through an increase on energy bills (from which low-income sectors are exempt).

The operation of the energy facilities depends mostly on private funding.[44] Also, as mentioned above, the Brazilian National Development Bank (BNDES) has special financing available for PROINFA projects (up to 70% of capital costs, excluding site acquisition and imported goods and services).[45]

2. Personnel Training

The Chamber of Commerce, where the commercial energy transactions occur, promotes the training of every participant in this market: energy generators; operators and officials; government energy personnel; energy distribution facilities; energy consumers (that do not have energy provided by energy distribution facilities); and energy dealers. Such training guides the economic participants on how to play in the energy market, especially regarding market rules.[46]

---

[43]   Governo Federal, Comitê Interministerial sobre Mudança so Clima, Plano Nacional sobre Mudança do Clima, PNMC [National Plan for Climate Change], Dez. 2008, P35, Ministério do Meio Ambiente, http://www.mma.gov.br/estruturas/169/_arquivos/169_29092008073244.pdf (last accessed 9 August 2011).

[44]   Ibid.

[45]   Programme of Incentives for Alternative Electricity Sources (PROINFA), World Resources Institute, http://projects.wri.org/sd-pams-database/brazil/programme-incentives-alternative-electricity-sources-proinfa (last accessed 13 April 13 2013).

[46]   Câmara de Comercialização de Energia Elétrica, CCEE, Visão Geral das Operações na CCEE [General Scenario of Deals Celebrated inside the Chamber of Commerce of Eletric Energy], Versão 2011, p. 96, http://www.ccee.org.br/StaticFile/Arquivo/biblioteca_virtual/Visao_Geral_das_Operacoes_CCEE_2011.pdf (last accessed 9 August 2011).

The commercialization of carbon credits generated by PROINFA also required the training of government personnel for establishment of the Project Design Document to be sent to the Clean Development Mechanism Executive Board.[47]

3. Public Participation Provisions

The program receives contributions from the public through participation of private agents in energy tenders; public hearings during the environmental impact assessment are required as a condition for environmental licenses of energy facilities;[48] and public hearings also are required prior to decisions that may affect the rights of the participants in the energy sector or of the consumers with respect to laws and regulations.[49]

4. Environmental Assessment

Power plants are required by law[50] to do an environmental impact assessment prior to the construction of any facility that may cause significant impact on the environment in order to be granted environmental permits.

5. Net Savings or Costs

PROINFA is expected to create 150,000 jobs,[51] especially in the Northeast region, which is one of the less developed areas of the country (its Human Development Index (HDI) was 0.720 in 2005,[52] when the country's HDI was 0.792[53]).

The total costs for the implementation of the program are expected to be R$ 10.14 billion, equivalent to U$ 5.359 billion. From this amount,

---

[47] Eletrobrás, 'Plano Anual do Proinfa 2010' [Proinfa Annual Plan 2010], Agência Nacional de Energia Elétrica, ANEEL, http://www.aneel.gov.br/cedoc/areh2010930_3.pdf (last accessed 9 August 2011).
[48] Federal Constitution, article 225, §1, IV and Resolução CONAMA article 11, §2.
[49] Law n 9.427/1996, article 4, §3.
[50] Federal Constitution article 225, §1, IV and Resolução CONAMA 001/1986 article 2.
[51] Eletrobrás, http://www.Eletrobrás.com/elb/Proinfa/data/Pages/LUMISA BB61D26PTBRIE.htm (last accessed 9 August 2011).
[52] PNUD, http://www.pnud.org.br/pobreza_desigualdade/reportagens/index.php?id01=3039&lay=pde (last accessed 9 August 2011).
[53] PNUD, http://www.pnud.org.br/pobreza_desigualdade/reportagens/index.php?id01=1445&lay=pde (last accessed 9 August 2011).

R$ 7 billion (U$ 3.7 billion) are expected to be funded by federal banks.[54] The annual income of PROINFA is expected to be R$ 2 billion (U$ 1.057 billion).[55]

E. PROBLEMS AND SOLUTIONS

There is a great dispute between energy generators and the federal government regarding the title to carbon credits generated by the energy contracted under PROINFA. The Decree establishing PROINFA attributed to the federal government-owned company the title to the carbon credits generated under PROINFA.[56]

The constitutionality of this Decree was disputed because it exceeded what was established in the authorizing law by creating new rules and extinguishing rights which, in accordance with the Brazilian law, can only be done by law, and not by a decree.[57]

This very interesting dispute questioned whether the carbon credit should be entitled to the entity that generates the energy or to the entity that purchases such energy. Moreover: should the federal government have the right to own the carbon credits because it was buying the energy and facilitating the finance of the project? And if the credits were not delivered to Eletrobrás, should the amount payable under the PPA be reduced? Should the project developer have the right to the credits, as it is absorbing most of the risks? These were a few issues that are still without an answer.

The Brazilian company, Goiasa Goiatuba Álcool LTDA, filed suit for a

---

[54] BNDES (Banco Nacional do Desenvolvimento Econômico e Social), BASA (Banco da Amazônia), CEF (Caixa Econômica Federal), BB (Banco do Brasil) e BNB (Banco do Nordeste do Brasil). See Eletrobrás, http://www.Eletrobrás.com/elb/Proinfa/data/Pages/LUMISABB61D26PTBRIE.htm (last accessed 9 August 2011); see also Governo Federal, Comitê Interministerial sobre Mudança so Clima, Plano Nacional sobre Mudança do Clima, PNMC [National Plan for Climate Change], Dez. 2008, P35, Ministério do Meio Ambiente, http://www.mma.gov.br/estruturas/169/_arquivos/169_29092008073244.pdf (last accessed 9 August 2011).

[55] Eletrobrás, http://www.Eletrobrás.com/elb/Proinfa/data/Pages/LUMISAB B61D26PTBRIE.htm (last accessed 9 August 2011).

[56] Article 12, §2 of Decree n 5.025/04, with the text introduced by Decree n 58 Eletrobrás 82/06: 'Eletrobrás shall develop, directly or indirectly the processes of preparation and development of the Project Design Document (PDD) registry, monitoring and certification of the Emission Reductions and also the commercialization of the Certified Emission Reductions generated by the project under PROINFA.'

[57] Federal Constitution, article 5, II.

writ of mandamus in the Federal Supreme Court, arguing that it should be entitled to the carbon credits generated by the project. Eletrobrás argued that, based on Decree n 5.882/06 (which has questionable constitutionality), it was entitled to the carbon credits generated by any company under PROINFA, including Goiasa Goiatuba Álcool LTDA, and if Goiasa Goiatuba Álcool LTDA wanted to use the credits, it should offset the costs of the price agreed to by Eletrobrás.

Goiasa Goiatuba Álcool LTDA argued that the Decree is illegal, as the PROINFA law did not specify the entitlement of the carbon credits to Eletrobrás, but only the acquisition of energy.

Such problems caused an enormous legal uncertainty in the market, as the company executed the contract before the enactment of the Decree, harming rights and discouraging investments. Changing the rules in the middle of the game discourages investment of foreign capital in the country and that may be the consequence of the legal upholding of the Decree.

The company requested an injunction to hold Decree n 5.882/06 unconstitutional, and an order granting allowance to it of the carbon credits arising from the company operations under PROINFA, without any price reduction.

Besides the dispute regarding the title to carbon credits, PROINFA faces other sorts of uncertainties. The program establishes goals to be achieved in the long term, but the non-compliance with such goals erodes the legal predictability by potential investors that they need in order to feel secure to invest the large amount of money that these kinds of projects require.

PROINFA's actual system of production and marketing of renewable energy sources creates some obstacles for their achievement of a better position in the energy market. First, the difference between the price of conventional energy and the price of alternative energy sources is subsidized. Therefore, the energy sources under PROINFA depend on the capacity of the government's investment, which is made through the Brazilian National Development Bank (BNDES) to ensure production increases or keeps to the same levels. The dependency on this financing source is a limiting factor to project expansion and it makes financing very vulnerable to the investment power and politics of the government.

Second, the PROINFA program provides no legal warranty ensuring demand or a decent price (considering the investments made by the investors) for the energy generated by biomass, small hydro or wind exceeding the 3,299.40 MW covered by the program's First Phase. That said, the generation of renewable energy that goes beyond this level can generate losses to the producers, as they will not be able to count on PROINFA's benefits and will suffer competition from the conventional sources in negotiations in the competitive market. This reality takes from the investors the incentive to expand production beyond the level established by the government.

Third, PROINFA's price warranty encourages production inefficiency because the producer does not feel the need to invest in technological innovation and efficiency improvements. Considering these terms, the assurance of the selling price contributes to the maintenance of higher prices, highlighting the dependency of such sources on the government subsidies of PROINFA. The observation of the PROINFA wind energy price in comparison with the first energy tender exclusive to wind sources, which was made outside PROINFA, shows that when alternative energy is sold in the free market, the selling value is much lower than the one provided by PROINFA.

Fourth, the PROINFA program does not internalize the negative externalities (social and environmental damage) generated by conventional sources, which is a major problem of the energy sector as a whole. Therefore, conventional energy is still considered cheaper and continues to be the main government option when decisions are made regarding energy expansion planning for supporting national economic growth.

However, it is important not to understate PROINFA's merit. The program was essential to the introduction of the alternative energy sources to the Brazilian grid and in the creation of a market (still small) to sell renewable energy. Furthermore, the PROINFA mechanism inserts in the government's agenda the necessity to reconcile energy generation with environmental conservation.

F. CONCLUSION

The First Phase of PROINFA was predicted to finish by the end of 2011. The government still is silent about the continuation of the program and some government officials believe that the program will not continue because it already has fulfilled its main objective: the creation of offer and demand of new renewables.[58] In the future, it is expected that the country will keep the tax deductions and other incentives and subsidies to modern renewables that are already in place in order to make them competitive to

---

[58] Elisângela Mendonça, 'Adeus, Proinfa 2' [Goodbye, Proinfa 2], 2 fev. 2010, Brasil Energia, http://www.energiahoje.com/brasilenergia/notici ario/2010/02/02/403330/adeus-proinfa-2.html, (last accessed 9 August 2011).

other sources of energy, thereby stimulating their acquisition in regular energy tenders.

1. Light for All Program

PROINFA does not include solar energy because this source was considered to be still too expensive to be included in the grid base load. The Brazilian government opted for using solar energy where such source is economically more competitive, in off-grid communities. It is estimated that in Brazil the generation in off-grid systems costs R$ 800–1,200/MWh (U$ 462–693/MWh) which is 4 to 5 times more expensive than other grid generation.[59]

In order to supply off-grid communities with electricity, the federal government launched the 'Light for All Program.' This program is aimed at electrifying all rural areas of the country by extending the grid or by generating electricity through decentralized regimes, such as with photovoltaic panels.[60]

The generation of electricity through decentralized systems was adopted for supplying electricity to low density communities, especially in the Amazon region. The program seeks to install 348 generation units in the North and Northeast regions of the country, at a predicted cost of R$ 8.4 million (U$ 4.8 million).[61]

One of the energy generation projects created within the Light for All Program, the pilot project Xapuri, received technical and financial aid from the German government, through a technical cooperation agreement

---

[59] Centrais Elétricas Brasileiras S.A., Eletrobrás, Inter-American Institute for Cooperation on Agriculture, 'Projeto de Cooperação Técnica para Acesso e Uso da Energia Elétrica como fator de desenvolvimento de comunidades do meio rural brasileiro' [Project of Technical Cooperation for Access and Use of Electric Energy as a development instrument for communities in Brazilian rural areas], BRA/IICA/09/001 Relatório [Report] n 06/07. 36 (2011), Inter-American Institute for Cooperation on Agriculture, http://www.iica.int/Esp/regiones/sur/brasil/Lists/DocumentosTecnicosAbertos/Attachments/316/Jos%C3%A9%20Carlos%20Vilela%20Ribeiro%20-%2020110127%20-%20eletrobr%C3%A1s.pdf (last accessed 1 August 2011).

[60] Decreto n 4.873/2003, article 6.

[61] Marta Olivieri, 'Experiência da Eletrobrás com Projetos e Implantação de Minirredes' [The Experience of Eletrobrás regarding Projects and Implementation of Small Grids], [Small Grids and Hibrid Systems associated with Renewable Energy in Rural Electrification], Minirredes e sistemas híbridos com energias renováveis na eletrificação rural,' 26 May 2011, Laboratório de Sistemas Voltaicos. http://lsf.iee.usp.br/lsf/index.php?option=com_content&view=article&id=52&Itemid=68 (last accessed 1 August 2011).

established between Brazil and Germany in 2005.[62] The cooperation agreement was supposed to finish in 2008, but it was extended for an undetermined period of time.[63]

The Xapuri project aims to develop a technical and institutional model of exploitation of solar energy guided by the following principles: universalization of electrical supply; stimulation of development in the poorest areas; and protection of the environment.[64] The project acts on three fronts: it explores which technology is most appropriate to generate energy for Amazon communities; it evaluates the existing norms applicable to decentralized generation and it proposes normative improvements; and it trains the personnel from local electricity distribution utilities in order to enable them to replicate the experience in other places.[65]

The solar panels installed in each unit were predicted to supply enough energy to provide each residence with the following services:[66]

- Illumination by three bulb lamps, 600 lumens each, operating 4 hours/day
- TV and DVD operating 4 hours/day
- Recharge of a mobile phone

---

[62] Federal Government, Divisão de Atos Internacionais, Ajuste Complementar, ao Acordo Básico de Cooperação Técnica, sobre O Projeto 'Ações para Disseminação de Fontes Renováveis de Energia em Áreas Rurais no Norte e Nordeste do Brasil' [Complementary adjust to the Technical Cooperation Agreement regarding the Project 'Actions to Disseminate Renewable Energy in Rural Areas of the North and Northeast Region of Brazil'], Pn: 2001.2511.2, Ministério das Relações Exteriores, http://www2.mre.gov.br/dai/b_rfa_506_5531.htm (last accessed 1 August 2011).

[63] Federal Government, Divisão de Atos Internacionais, Ajuste Complementar, por Troca de Notas, 'ao Acordo Básico de Cooperação Técnica entre o Brasil e a Alemanha sobre a Continuidade de Projetos de Cooperação Técnica' [Complementary adjust by exchange of notes to the Basic Technical Cooperation Agreement by and between Brazil and Germany regarding the continuity of Projects and Technical Cooperation], Ministério das Relações Exteriores, http://www2.mre.gov.br/dai/b_alem_520.htm (last accessed 1 August 2011).

[64] Eletrobrás, 'Projeto de Cooperação Técnica Eletrobrás/GTZ' [Technical Cooperation Project Eletrobras/GTZ], http://www.Eletrobrás.com/elb/main.asp?Team=%7B565E0DFC-FD35-456D-A46F-DAB3DD876DFE%7D (last accessed 1 August 2011).

[65] Ibid.

[66] Eletrobrás, Relatório [Report] n 06/07, 11, 2011, Inter-American Institute for Cooperation on Agriculture, http://www.iica.int/Esp/regiones/sur/brasil/Lists/DocumentosTecnicosAbertos/Attachments/316/Jos%C3%A9%20Carlos%20Vilela%20Ribeiro%20-%20110127%20-%20eletrobr%C3%A1s.pdf (last accessed 1 August 2011).

- Refrigerator of 150 liters operating between 3 and 7 degrees Celsius
- Food blender operating 5 minutes/day
- Fan operating 1 hour/day

After the installation of the solar panels, the results were monitored for 18 months.[67] The evaluation of the project assessed the following aspects: technical performance of the chosen photovoltaic system; quality of the service provided; customers' satisfaction; financial sustainability of the project; social-economic impact of the project on the users' quality of life; and the effect on the micro-regional development and on the local market of domestic appliances.[68]

The analysis of monitoring results stressed the importance of: investing in energy efficiency in off-grid communities because of the high cost of energy in such places; and regulating the use of micro-grids to connect the small size generation units.[69]

2. Proálcool

a. Brief project description
The National Alcohol Program (Proalcohol), (in Portuguese: Programa Nacional do Álcool – Proálcool) was created in November 1975 by Decree n 76.593 and it was the first Brazilian program to incentivize the use of alternative sources of energy. The program was launched due to the oil crises in the 1970s, which raised the price of oil and its derivatives in the international market, and it was not guided by a sustainability concern.

b. Why initiated
The program was launched by the federal government in a glorious time in Brazilian history known as the 'Brazilian Economic Miracle' and it was

---

[67]   Eletrobrás, 'Projeto de Cooperação Técnica Eletrobràs/GTZ' [Technical Cooperation Project Eletrobras/GTZ], http://www.Eletrobràs.com/elb/main. asp?Team=%7B565E0DFC-FD35-456D-A46F-DAB3DD876DFE%7D   (last accessed 1 August 2011).

[68]   Eduardo Borges et al., 'Sistemas Fotovoltaicos Domiciliares' [Residential Photovoltaic Systems], Eletrobrás, http://www.Eletrobràs.com/elb/main.asp? Team=%7B565E0DFC-FD35-456D-A46F-DAB3DD876DFE%7D (last accessed 1 August 2011).

[69]   Eletrobrás, Relatório [Report] n 06/07, 35-37 (2011), Inter-American Institute for Cooperation on Agriculture, http://www.iica.int/Esp/regiones/sur/ brasil/Lists/DocumentosTecnicosAbertos/Attachments/316/Jos%C3%A9%20 Carlos%20Vilela%20   Ribeiro%20-%20110127%20-%20eletrobr%C3%A1s.pdf (last accessed 1 August 2011).

established to encourage the use of ethanol in gasoline, using the maximum extent feasible, in existing vehicles (approximately 20% by volume).[70]

In 1979[71] the program was boosted as the government was trying to find solutions to develop new sources of renewable energy.

Proálcool was managed by the Ministry of Industry and Commerce (in Portuguese: Ministério da Indústria e Comércio) through a National Executive Commission of Ethanol (in Portuguese: CENAL, Comissão Executiva Nacional do Álcool) with the objective to produce ethanol to replace or reduce the consumption of oil and its derivatives, especially gasoline.

The first goal was to reach 3 billion liters of ethanol in 1980. In 1974 and 1975, before the program, the production was 625 million liters.[72]

Initially, the ethanol program promoted the addition of ethanol to gasoline for use in motor vehicles, and in 1977 the government encouraged the use of vehicles totally fuelled by ethanol.

The Brazilian economy at that time was threatened by the growth of the country's external debts, which increased from US$ 12.5 billion in 1973 to US$ 29 billion in 1976,[73] and in response to these external debts, the government decided to increase reliance on national investments in ethanol to boost the economy.

The ethanol program had an important role to play in this context because 80 percent of the oil consumed in Brazil was imported[74] and the importation of oil represented 50 percent of the country's total importation costs in 1979.[75]

During the period between 1973 and 1979, the government was trying

---

[70] University of Campinas, http://www.inovacao.unicamp.br/etanol/report/inte-etanol060710.pdf (last accessed 1 August 2011).

[71] http://brasilbiocom.wordpress.com/2009/03/20/a-crise-mundial-do-petro leo-e-o-proalcool-no-brasil/ (last accessed on 1 August 2011).

[72] Fernando H. de Melo and Eduardo G. da Fonseca, 1981, 'Proálcool, Energia e Transporte' [Proalcohol, Energy and Transportation], São Paulo: Pioneira/FIPE.

[73] Veja Online, Arquivo Veja [Veja File], 'No ar, o programa de racionaliza-ção' [On air, the rationalization program] (January 1977), http://veja.abril.com.br/idade/exclusivo/petroleo/190177.html (last accessed 28 October 2011).

[74] Pery F.A. Shikida and Carlos José C. Bacha, 'Evolução da Agroindústria Canavieira Brasileira de 1975 a 1995' [Evolution of the Brazilian Sugar Cane Agribusiness from 1975 to 1995], 2, Biblioteca Digital FGV (Fundação Getulio Vargas) http://bibliotecadigital.fgv.br/ojs/index.php/rbe/article/down load/746/1740 (last accessed 28 October 2011).

[75] Veja Online, Arquivo Veja [Veja File], 'No ar, o programa de racionaliza-ção' [On air, the Rationalization Program], (January 1977), http://veja.abril.com.br/idade/exclusivo/petroleo/190177.html (last accessed 28 October 2011).

to reduce the consumption of gasoline. Towards this end, the federal and local governments set in place several measures to discourage the use of individual vehicles in order to reduce the consumption of gasoline.[76] On one hand such measures were successful in reducing the consumption of gasoline by 10 percent; on the other hand, this policy was not successful in reducing the importation of oil because the country did not reduce consumption of other oil derivatives, such as natural gas and diesel fuels.

Therefore, Brazil continued its need to refine the same amount of oil, and the final result was the need to export the gasoline[77] that was saved. The ethanol program only acquired national importance in 1979, when the federal government decided to invest USD 5 billion on subsidies to alcohol-run cars and to companies that produced ethanol in order to create demand and supply for the product.[78] The price of the alcohol was kept 30 percent cheaper than gasoline.[79] The results of these measures were impressive: the proportion of ethanol-run automobiles rose from 0.46 percent in 1979 to 76.1 percent in 1986.[80]

During the early 1980s, the ethanol program was doing really well and ethanol made up roughly half of Brazil's liquid fuel supply.

In 1985 oil prices started to decrease and, despite this, ethanol consumption in the domestic market did not change due to the artificial maintenance of the ethanol price (lower than the price of gasoline and diesel) and the reduced taxes on ethanol-run automobiles.

This situation remained until 1988, when the Brazilian government's shortage of resources to grant subsidies prevented the sugar cane industry from meeting the growing demand for ethanol at the same prices maintained until then. The maintenance of low prices paid to producers during the following ten years has contributed to the reduction in ethanol production, and the importation of foreign (gasoline-run) cars has contributed to a further reduction in ethanol demand.

At that time, tests demonstrated that by adding ethanol to gasoline it was possible to extract heavy metals, such as lead, from gasoline, thus maintaining fuel quality and engine performance. This measure was

---

[76] Ibid.
[77] Veja Online, Arquivo Veja [Veja File], 'O petróleo da cana' [The Sugar Cane Oil], (June 1979), http://veja.abril.com.br/idade/exclusivo/petroleo/130679.html (last accessed 28 October 2011).
[78] Ibid.
[79] Ibid.
[80] Biodieslbr, 'Proálcool – Programa Brasileiro de Álcool' [Proalcohol – National Alcohol Program], http://www.Biodieslbr.com/proalcool/pro-alcool.htm (last accessed 28 October 2009).

responsible for a considerable improvement in the air quality in larger cities.[81] The success of this mixture has prompted European countries to adopt similar measures, motivated mainly by the reduction in fossil fuel consumption by replacing it with biofuels.

The mix of ethanol and gasoline has guaranteed the maintenance of the ethanol demand, disassociating it from the state subsidies that are subject to variations in the government's investment capacity.

At the end of the 1990s, there was the deregulation of the sugar cane industry and the liberalization of alcohol prices, which became ruled by the market law of supply and demand. Such facts contributed to the selection of the sugar cane industry actors, which resulted in an enhancement of the productivity and efficiency of the sector.[82]

In 2003 nationally developed 'flex' cars were introduced in the market that could use any mixture of gasoline and ethanol, including pure ethanol, and this created a new boom in the consumption of ethanol.[83] Flex-fuel vehicles were greatly accepted by the population and in the second half of the 2000s they represented 80 percent of the fleet of the country. This development significantly advanced the Brazilian ethanol program.

G. PUBLIC/PRIVATE FUNDING AND PARTICIPATION BY FUNDERS AND/OR INTERNATIONAL ORGANIZATIONS

The greatest tool used by the government to stimulate sugar cane production and industrial ethanol capacity was the subsidies to approved projects under the program.

The ethanol program was created by the government and the use of ethanol was promoted by offering credit at a low interest, assuring the purchase of the ethanol for reasonable prices and that ethanol had a better price than gasoline.

---

[81] José Goldemberg and Oswaldo Lucon, 2008, 'Energia, meio ambiente e desenvolvimento' [Energy, environment and development], 3rd ed. São Paulo: Edusp.

[82] Luis Fernando Paulillo et al., 'Álcool combustível e biodiesel no Brasil: quo vadis?' [Alcohol Fuel and biodiesel in Brazil: quo vadis?], Rev. Econ. Sociol. Rural, vol.45, n 3, Brasília, July/September 2007, http://www.scielo.br/scielo.php?script=sci_arttext&pid=S0103-20032007000300001&lang=pt (last accessed 1 August 2011).

[83] Biodieslbr, 'Proálcool – Programa Brasileiro de Álcool' [Proalcohol – National Alcohol Program], http://www.Biodieslbr.com/proalcool/pro-alcool.htm, (last accessed 11 July 2009).

It is important to highlight that Petrobras (the Brazilian oil company) made a decision to invest in distribution of ethanol as well, and the result was that between 1975 and 1979 ethanol production increased over 500 percent.[84]

In the early implementation phase, the ethanol program had to face public skepticism. In order to prove the reliability of the new technology, the example given by the state-owned telephony company, Telesp, which converted all its fleet to alcohol-run vehicles, was of substantial importance. Besides this measure, the government also developed several publicity campaigns to inform the population of the benefits of adopting alcohol-run cars.[85]

The previous policy of reducing the use of gasoline also played a role in convincing the population to adopt the new fuel: the public perception was that citizens had to choose between adopting ethanol and giving up on having an individual car. Such interpretation is comprehensible considering that the country was under a dictatorship (from 1964 to 1985), in which governmental interferences with the private sector were frequent and intense.

During the 1970s, the federal government awarded companies engaged in ethanol production with fiscal incentives,[86] in order to stimulate the ethanol production to supply the growing demand. In 1979 due to other oil crises, the government started to subsidize the price of ethanol and to reduce taxation of ethanol-run cars.[87]

Since 2001, alcohol prices have been kept competitive in comparison to gasoline by the imposition of the Contribution for Intervening on the Economic Domain on fuels (in Portuguese: Contribuição de Intervenção no Domínio Econômico, Combustíveis (CIDE Combustíveis)). The Brazilian CIDE is a federal tax modality aimed at encouraging or discouraging a conduct and raising funds for mitigation measures related to the impacts caused by such conduct.[88] The CIDE Combustíveis is the

---

[84]   University of Campinas, http://www.inovacao.unicamp.br/etanol/report/inte-etanol060710.pdf (last accessed 1 August 2011).

[85]   Veja Online, Arquivo Veja [Veja File], 'O petróleo da cana' [The Sugar Cane Oil] (June 1979), http://veja.abril.com.br/idade/exclusivo/petroleo/130679.html (last accessed 29 July 2011).

[86]   Biodieslbr, 'Proálcool – Programa Brasileiro de Álcool' [Proalcohol – National Alcohol Program], http://www.Biodieslbr.com/proalcool/pro-alcool.htm (last accessed 11 July 2009).

[87]   Ibid.

[88]   Flávia Helena Gomes, 'Das implicações tributárias do desvio de finalidade na destinação do produto arrecadado com a CIDE-Combustíveis' [Tax consequences of the purpose deviation of the destination of resources collected by the

CIDE incident on the importation and commercialization of alcohol, oil and its derivatives. The CIDE incident on fuels is aimed at reducing the use of auto-vehicles and mitigating the environmental impacts caused by fuel production, transport and consumption.[89] The CIDE Combustíveis encourages the consumption of alcohol by charging smaller amounts on this fuel than from oil derivatives.[90] Despite the government incentive provided through the CIDE, since 2000 the expansion of alcohol production is determined by private initiatives, not by the government.[91]

The ethanol industries are also required by law[92] to carry out an environmental impact assessment prior to the implementation of the project in order for environmental permits to be issued.

H. PROBLEMS AND SOLUTIONS

As the ethanol program evolved some problems were evidently affecting public policies, energy, industry, agribusiness, transportation, exportation and importation, social issues, labor issues and environmental issues. It is hard to say that the ethanol program has not had problems as well as successes.

A lot of social problems were faced by the workers in the sugar cane fields, such as labor conditions, salaries, and even slavery. But many of these problems were fixed by government and police supervision.

Another important issue faced was waste water residues from the fields; but such residues can be used nowadays to fuel the boilers and generate energy, and as a fertilizer for farms.[93]

CIDE-Fuels' fee], in Revista do Direito Público da Universidade Estadual de Londrina, ano 3, vol. 3, set./dez. 2008.

[89] Law n 10,336/2001, article 1, §1. See also Flaviana Marques de Azevedo and Cíntia Bezerra de Melo Pereira Nunes, 'Tributos ambientais: uma análise da CIDE-Combustíveis incidente nas atividades desenvolvidas pela indústria do petróleo e do gás natural' [Environmental tax: an analysis of CIDE-Fuels' fee charged from activities developed by oil and natural gas companies], in Revista Direito E-Nergia, vol. 2, n 1, jan./jul. 2010.

[90] The CIDE charges R$ 860.00 per m^3 of gasoline and R$ 37.20 per m^3 of alcohol. Law n 10,336/2001, article 1, I and VIII.

[91] Biodieslbr, 'Proálcool – Programa Brasileiro de Álcool' [Proalcohol – National Alcohol Program], http://www.Biodieslbr.com/proalcool/pro-alcool.htm (last accessed 11 July 2009).

[92] Federal Constitution article 225, §1, IV and Resolução CONAMA 001/1986 article 2.

[93] Novacana.com (Biodieslbr), http://www.Biodieslbr.com/proalcool/proal-cool-externalidades.htm (last accessed 11 July 2009).

There are a lot of externalities to be considered including the environmental and social-economic impacts related to ethanol production, although ethanol was able to solve issues such as the elimination of lead in gasoline and reduction of pollution in the big urban cities and reduction of greenhouse gas emissions.[94]

I. CONCLUSION

Brazil still needs to expand its sugar cane production, estimated to increase from 420 (2006/07) to 1,140 million tons per year (2015/16)[95] and must do so without destruction of Amazon rainforests, exploitation of small landholders, and while meeting other environmental and fair labor challenges. It is important to highlight the necessity of infrastructure in the logistics of the transportation of ethanol due to the fact that its production usually occurs in remote areas. Despite all the historical problems, the program is considered a great success.

J. LESSONS LEARNED

In the mid-1990s Brazil privatized its energy sector and divided its energy regime into regulated and free market sectors. The regulated sector was required to purchase all the renewable energy offered to it through tenders from the Program of Incentives for Alternative Energy Sources (PROINFA), while the free market is merely encouraged to purchase renewable energy resources through various incentives, but may also purchase from PROINFA. PROINFA was expected to add 3,300 MW of renewable energy through 2011. Tenders were issued for 8,837 MW by 2011, of which 5,419 MW was added to the grid by 2010. So, the PROINFA mechanism was successful in exceeding its goals, but the amount of renewable energy installed and tendered was still a relatively small amount of Brazil's total electricity generation demand.

The government subsidizes the purchases from PROINFA by guaranteeing purchases of its power over a 20-year period by the government-owned utility, Eletrobrás. The prices paid are based on the government-determined value and cost of the resources. The Brazil

[94] Ibid.
[95] Instituto de Economia Agricola, http://www.iea.sp.gov.br/out/verTexto.php?codTexto=7448 (last accessed 11 August 2011).

National Bank of Economic and Social Development also subsidizes autonomous independent producers (PIAs) that are small renewable energy companies that are the primary deliverers of renewable energy. The program was initially financed in part by the sale of carbon credits under the CDM program of the Kyoto Protocol. A problem with these heavy subsidies is that they are dependent on continued government support, creating an uncertainty making it more difficult to raise private capital.

The mission of PROINFA is to guarantee the economic and financial feasibility of renewable resources. For example, it allowed Brazil in three years to increase its wind farm installed capacity from 22 MW to 414 MW. PROINFA is predicted to reduce 3 million tons of carbon dioxide emissions annually when it is fully implemented. It is projected to create 150,000 jobs, particularly in the economically depressed Northeast region. Also on the plus side, there has been extensive sharing in implementation of the project with the private sector, with public/private partnerships created. There were full public hearings on the project with transparency of the costs and benefits. Two full environmental impact statements were issued, a second when the scope of the project was expanded, and there was full public participation in the process. Also, the Chamber of Commerce conducted an extensive training program for all participants.

On the other hand, there is a long list of conditions in the renewable energy law for qualification for Eletrobrás mandatory purchases and bank subsidies. One provision requiring that PROINFA projects contain 60 percent Brazilian machinery and materials has been difficult to meet and is similar to a provision of Ontario (Canada) law, which has been challenged by Japan under the World Trade Organization regulations.

Also, there is an ongoing dispute as to which agency is entitled to the carbon credits generated from the program, claimed by the federal government, Eletrobrás and the PIA generators. This dispute is presently in litigation. The dispute creates uncertainty that discourages private investment in renewable energy projects.

The PROINFA system thus has introduced some new renewable energy into the Brazilian electricity generation system, but the amounts still are relatively small and costs relatively high. So the efficacy of this complex program in promoting renewable energy for electricity generation is not yet established. It certainly has not been as successful as the relatively simple feed-in tariffs in Germany and other countries.

1. Solar Energy

PROINFA did not include solar energy for the grid because at the time of its establishment this resource was considered too expensive for grid base load application. Even for off-grid applications solar was estimated to be 4 to 5 times more expensive than other grid options, but serving poor rural areas was considered by the government to be an important political and social necessity and solar was relatively economic for such off-grid applications. Brazil therefore launched its Light for All Program to serve rural communities for which grid electricity was unavailable.

The program's declared mission was to provide a universal electricity supply to meet the very basic needs of rural areas, to stimulate development in the poorest areas, and to protect the local environment. It assesses the options for decentralized generation to these communities, fosters the creation of local distribution utilities and trains its personnel, seeking to qualify staff capable of extending the program to other areas. The requisites for service are very basic: illumination by three light bulbs and TV/DVD service per house for 4 hours per day, recharge of a mobile phone, basic refrigeration, a good blender for 5 minutes/day and a fan operating for 1 hour/day. An evaluation of the project was conducted on the effectiveness of the program, concluding the importance to the community of even such basic services for off-grid communities and stressing the need to regulate the use of micro-grids to connect such small size generation units. Based on this initial evaluation, the project has been continued.

A unique innovation for serving rural areas with biodiesel is a 'Social Seal' program that gives tax incentives to biodiesel developers who devote a given percentage of their investment to poor rural areas. While the program has not been implemented very efficiently in Brazil it is a model well worth exploring for other programs.

2. Sugar Cane Ethanol Program for Transportation (Proálcool)

This program was initiated in 1975 at a time of soaring gasoline costs to save money for the populace and the country, and reduce the pollution from gasoline propulsion. In these respects it was very successful. Ethanol was initially subsidized by the government but within a few years was successful enough to permit elimination of the subsidies. At first ethanol was required to be blended with gasoline, but in 1977 the government encouraged the use of vehicles totally fueled by ethanol and automobile fleets were required to acquire flex-fuel vehicles capable of using any amount of this fuel.

Prior to the program, Brazil imported 80 percent of its oil use, representing 50 percent of its very high importation debts. The program was successful in reducing the consumption of gasoline by 10 percent from 1973–79, but was not successful in reducing oil imports because of the need for continued imports to supply diesel fuel for trucks, thus creating a surplus of gasoline that Brazil had to export. The lesson here is for the need of more thorough planning. Brazil had to provide subsidies for ethanol, keeping its price 30 percent lower than gasoline and raising the proportion of ethanol cars from 46 percent in 1979 to 76 percent in 1986. The program was very successful in reducing air pollution, permitting elimination of lead from gasoline, improving vehicle performance and improving the air quality in cities. Flex-fuel vehicles capable of using any proportion of ethanol were produced locally and were very popular; by the late 2000s they represented 80 percent of the country's vehicles.

On the downside, the early years of the ethanol program had lots of environmental and social problems. Valuable tropical forests in the Amazon were cut down to accommodate sugar cane production; sugar cane monocultures were planted, destroying the productivity of the land and sustainability of the project; lands used by indigenous people over the years to which there was no proof of title were seized by large local and international producers; profits from ethanol operations were taken by the large local and international producers who failed to pay local communities fair compensation for the land taken; sub-standard wages, inadequate housing and miserable working conditions were imposed on sugar cane workers; no health care provisions were made for workers in the very hazardous work of harvesting sugar cane, and the fields were burned after harvesting, causing serious pollution. Again, poor planning is to be faulted for permitting these conditions to develop.

As these abuses were exposed, laws were passed to address them, but to this date these laws are inadequately enforced. There are too few enforcers, they are underpaid, inviting corruption, and they are undertrained. Nevertheless, progress is being made to improve the conditions of the program concerning the protection of property rights, working conditions, prohibition of deforestation in the Amazon, requirements for crop rotations every three years, etc. The processing of sugar cane to produce the ethanol has become very efficient, using the sugar cane stalks to energize the refineries, for example.

Much of this progress is attributable to the work of an international voluntary organization called the Roundtable for Sustainable Biofuels which, like the Roundtable for Sustainable Palm Oil, has created standards for sustainable production and a program for certification of sustainably grown products. Both organizations are having success in persuading purchasers of biofuels and diesel fuels to purchase only products certified as sustainable. These are very important innovative lessons, and perhaps the only effective way of combatting corruption in the producing countries. If the purchasing countries will not buy the products that avoid sustainable practices through corruption, there will be internal pressures to prevent the corruption. The European Union countries, large ethanol and diesel oil purchasers, have been the leaders in effectuating the purchase restrictions of buying only products certified as sustainable.

Brazil has become the world's largest producer of ethanol and the number one exporter to all countries except the United States, which puts a large tariff in place to protect its domestic production of ethanol from corn that is a feedstock far inferior to sugar cane in production of ethanol and has food supply and price problems with its use. Ethanol is one of Brazil's most profitable sources of export income.

At the bottom line, ethanol and palm oil have great potential to relieve the world's dependence on fossil fuels and reduce the emission of greenhouse gasses, but only if strict and well-enforced regulations are implemented to prevent destructive practices. The example of Indonesia's destruction of its forests and greenhouse gas-absorbing peat bogs is illustrated in Chapter 7. Brazil is on its way to overcoming the environmental and social obstacles to sustainable development of its biofuels.

7. Case study of Indonesia's palm oil-based biodiesel program

Richard L. Ottinger with Christopher J. Riti

A. PROJECT DESCRIPTION

Over the past several decades, Indonesia has transformed itself into the world's leading producer of crude palm oil (CPO), producing over 18 million tons annually.[1] Capitalizing upon relatively inexpensive labor, abundant land, and a favorable regulatory environment, the production and export of palm oil and its derivative products have come to constitute a sizeable share of the national economy.[2] Exports have increased by 274 percent over the past decade,[3] and the industry is expected to continue expanding.[4] Palm oil is currently the world's number one source of vegetable cooking oil, but is also incorporated into countless processed foods and other household products, from cookies and candies to cosmetics, shampoos, and industrial lubricants.[5]

---

[1] World Growth, 'The Economic Benefit of Palm Oil to Indonesia' 4 (2011).

[2] See, e.g., Colin Barlow et al., 'The Indonesian Oil Palm Industry' 2 (2003) (in the early 2000s, palm products 'contributed 1.5–2% of the country's gross domestic product'). Moreover, approximately 1.3 million hectares (ha) of new land in Indonesia were dedicated to palm oil since 2005, reaching almost 5 million ha in 2007. This figure represents about 10.3% of the 48.1 million ha apportioned to agricultural lands. See World Growth, supra note 1, at 11.

[3] Emerging Markets Direct, 'Indonesia Palm Oil Industry – Overview, Trends, Prospects and Swot Analysis' (2010), http://www.articlesnatch.com/Article/Indonesia-Palm-Oil-Industry---Overview--Trends--Prospects-And-Swot-Analysis-/1875678 (accessed April 12, 2013) (estimating that Indonesia could satisfy 57% of annual growth in demand for palm oil).

[4] See, e.g., John Slette and Edy Wiyono, US Dept. of Agriculture, 'Indonesia Oilseeds and Products Annual 2–3' (2011) (USDA predicts a growth from 17.85 million metric tons (mmt) in 2010 to 19.35 mmt in market year (MY) 2011–12. By 2020, palm oil production and consumption globally will increase to 60 mmt, see, supra note 1, at 15.

[5] Fred Pearce, 'Sustainable Palm Oil: Rainforest Savior or Fig Leaf?' Yale

Palm oil may also be refined into biodiesel. Oil palms produce average yields five to nine times higher than other seed-bearing crops like soybean and rapeseed, thus rendering them an attractive feedstock in terms of both cost competitiveness and efficiency.[6] Despite the fact that palm oil currently comprises only about 5 percent of global biodiesel production, demand is projected to grow swiftly considering the level of interest in renewable energy sources.[7] The Indonesian government, recognizing both the value of such a commodity and the comparative advantages available, has actively encouraged further development of its oil palm plantations as well as domestic biofuel refining capacity. The government also endeavors to become the 'best sustainable palm oil producer in the world,' producing 40 million tons of palm oil by 2020, with 50 percent of that total devoted to bioenergy.[8] To realize this goal, the government has introduced biofuel production strategies and policies, with an intended target of 6 million tons of palm oil allocated to the processing industry each year.[9] The majority of this biodiesel is exported, mainly to Australia, the European Union, and the United States.[10]

Unfortunately, this interest in palm oil-based biofuels, and in palm oil generally, has had a disastrous impact on Indonesia's natural ecology. Over 18 million hectares of virgin forest have been cleared in the name of sustained palm oil plantation development.[11] Distressingly, only six of the 18 million hectares of forest that have been converted for oil palm have actually been planted with palm trees.[12] This trend reflects the understanding that oftentimes companies will procure conversion permits for nominal oil palm development, when in reality access to valuable (and

---

360, 29 Nov. 2010, http://e360.yale.edu/feature/sustainable_palm_oil_rainforest_savior_or_fig_leaf/2345/ (accessed April 12, 2013).

[6] World Growth, supra note 1, at 15.

[7] See ibid. at 21.

[8] World Bank Group, the World Bank Group Framework and the International Finance Corporation, 'Strategy for Engagement in the Palm Oil Sector' 13 (2011).

[9] World Growth, supra note 1, at 8; P. Thoenes, Food and Agriculture Organization of the United Nations Commodities & Trade Div., 'Biofuels and Commodity Markets – Palm Oil Focus 1' (2006).

[10] Foreign Agricultural Service, US Dept. of Agriculture, 'Indonesia Biofuels: Biofuel Annual 2008 (2008).

[11] See, Marcus Colchester et al., 'Promised Land, Palm Oil and Land Acquisition in Indonesia – Implications for Local Communities and Indigenous Peoples' 29 (2006).

[12] Ibid. at 14, 29.

discounted) timber is the true goal.[13] The World Bank estimated that around 40 percent of Indonesia's 'legal' timber supply was derived from land clearance for oil palm,[14] even despite a 1993 Ministry of Forestry decree that prohibited transferring cleared land until conversion to a palm plantation was complete.[15]

Given palm oil's dismal legacy of environmental destruction, the prospect of dramatically increased demand for renewable biofuels over the next several decades is truly worrisome. In 2009, the European Union issued a directive imposing a renewable energy fuel target for use in transport of 10 percent by 2020.[16] While sustainable criteria are proposed in relation to this target,[17] meeting this mandated demand through palm oil would require additional land investments of 4–5 million hectares.[18] Satisfying China's growing requirements would require an additional 1 million hectares.[19]

Ensuring the propagation of sustainable management techniques, dedicated conservation programs, and best industry practices is critical to the survival of the world's remaining forests. And Indonesia has a valuable interest in fostering positive environmental outcomes, as export growth to Europe, the United States, and other large markets is increasingly dependent upon compliance with sustainable standards of production including limits on deforestation.[20]

---

[13] See, e.g., ibid. at 11; Jan Willem Van Gelder, Friends of the Earth, 'Greasy Palms, European Buyers of Indonesian Palm' 19 (2004).

[14] Colchester et al., supra note 11, at 29.

[15] See ibid. at 67.

[16] European Union (2009) 'Directive on the Promotion of the Use of Energy from Renewable Sources' (2009/28/EC), http://eur-lex.europa.eu/LexUriServ/LexUriServ.do?uri=CELEX:32009L0028:EN:NOT (accessed April 15, 2013) (hereinafter EU Directive).

[17] See ibid. art. 17–19. See also Section D, infra, for a discussion of these sustainability criteria.

[18] See UNEP, 'Towards Sustainable Production and Use of Resources: Assessing Biofuels' 68 (2009); World Growth, supra note 1, at 8; Pearce, supra note 5 (palm oil is likely to require an extra 30,000–70,000 square kilometers in the coming decade).

[19] World Growth, supra note 1, at 8.

[20] Slette, supra note 4, at 2–3. See also Angela Dewan, 'Biodiesel Demand Fuelled by Policy, Not Oil Prices' Center for International Forest Research (CIFOR) 6 Apr. 2011, http://blog.cifor.org/2359/biodiesel-demand-fuelled-by-policy-not-oil-prices/ (accessed April 16, 2013).

B.   HOW INITIATED

The first oil palm plantations in Indonesia were established in 1911. By 1967, following the nationalization of the Dutch colonial plantations, the sector covered about 106,000 hectares.[21] Under President Suharto, the Indonesian government spent the next several decades stimulating the development of the national oil palm sector, through direct investments, private partnerships, and attractive incentive schemes.[22] With the goal of becoming the world's leading palm oil producer, the government officially reserved 5.5 million hectares of mainly forested lands for plantation conversion.[23] Subsequent analysis of government records indicates that this number was far higher, with 9.13 million hectares in the country's eastern provinces alone.[24] By 1995, private companies were applying for additional allocations of up to 20 million hectares of forested areas;[25] the country proceeded to top 5 million hectares of productive oil palm in 2005,[26] and 6.1 million hectares in 2006.[27]

With the palm oil industry rapidly expanding and the global economy demonstrating growing interest in renewable biofuels, the government of Indonesia set its sights on developing its own palm oil-based biofuels for both export and domestic consumption. Biodiesel production began in 2005 on a commercial scale after the government announced an official target of allocating 6 million tons of palm oil to biodiesel production annually.[28] As fuel prices climbed over the next two years, the government continued to spend large sums to achieve wider market penetration of palm oil biodiesel. By 2007, eight plants were producing palm oil biodiesel, with total capacity of 765,000 tons.[29] Indonesia has the potential to produce 4.7 million kiloliters per year from 2015–25.[30]

---

[21]   Jan Willem Van Gelder, Friends of the Earth, 'Greasy Palms: European Buyers of Indonesian Palm Oil' (2004). (State-run companies, known as Persoroan Terbatas Perkebunan (PTPs) were an integral part of this strategy.) See ibid.

[22]   Ibid. at 18–19.

[23]   Ibid. at 19.

[24]   Ibid.

[25]   Ibid.

[26]   Colchester et al., supra note 11, at 21.

[27]   Indonesian Palm Oil Board, 'Indonesian Palm Oil in Numbers' 4 (2007).

[28]   Thoenes, supra note 9, at 4.

[29]   World Bank Group, 'Environmental, Economic and Social Impacts of Oil Palm in Indonesia: A Synthesis of Opportunities and Challenges' 10 (2010) (preliminary working draft) (hereinafter WBG Synthesis).

[30]   Anasia Silviati, US Commercial Service, 'Indonesia: Renewable Energy Market' 4 (2006).

C.   GOVERNMENT INVOLVEMENT

The government has always demonstrated a serious interest in promoting the oil palm industry. The industry matured gradually over several phases of government involvement. Support has come in the form of concessionary credits for estate development;[31] export tariffs, domestic market obligations, and other regulations;[32] and subsidized labor.[33] Palm oil-based biodiesel began to attract similar enthusiasm in 2006. Biofuels 'were seen as a strong sector in which to invest because the plantation sector had remained robust,' even in the wake of the 1997 East Asian Financial Crisis.[34] Energy security,[35] rural economic development and poverty reduction, and new export potential were all motivating factors in this approach.

During 2006, the government began to intensify its efforts at stimulating the nascent biodiesel industry, with various policies, subsidies, and research initiatives introduced to stimulate development in a favorable investment climate. The National Energy Policy of 2006 called for liquid biofuels to meet at least 5 percent of domestic energy needs by 2025.[36] The government promoted biofuel development broadly through the National Team for Biofuel Development (NTFBD) and instructed provincial and local governments to support biofuels through land procurement deals.[37] The government aimed to create 500,000 hectares of oil palm plantations

---

[31]   See, e.g., Diana Chalil, 'An Empirical Analysis of Asymmetric Duopoly in the Indonesian Crude Palm Oil Industry' 26, Univ. Sydney (2008) (the author notes a number of the unattractive investment features of palm oil development that necessitate subsidization, including the 'long maturation period, the high investment requirement, and the perishability' of the palm fruit); Van Gelder, supra note 21, at 19 (subsidies 'were intended to help investors overcome risks and uncertainties associated with establishing estates involving smallholders').

[32]   See, e.g., Chalil, supra note 31, at 19, 25; Van Gelder, supra note 21, at 25 (describing the CPO export tax initiated in 1998).

[33]   See Section E, infra, for information about the national transmigration program.

[34]   See Global Subsidies Initiative, Intl. Inst. for Sustainable Development, 'Biofuels – At what Cost? Government Support for Ethanol and Biodiesel in Indonesia' 17 (2008) (hereinafter GSI).

[35]   Indeed, Indonesia in 2006 for the first time became a net importer of petroleum, which likely underscored the need for alternative sources of transport fuel. See ibid. at 11.

[36]   See Presidential Regulation No. 5/2006 on National Energy Policy (Jan. 2006).

[37]   Presidential Decree No. 10/2006 on 'Establishment of National Team for Biofuel Development' (July 2006).

per year for increased biodiesel production, and would be offering millions of (US) dollars in concessionary loans to attract investment.[38] President Susilo Bambang Yudhoyono pledged US$110 million in credit interest subsidies for producers to assist in oil palm cultivation for biofuels.[39] In addition, the government planned to construct four biodiesel plants, at a total cost of US$33 million, and was prepared to spend US$1 billion to further build the industry.[40] Numerous other incentive packages were subsequently announced.[41]

By 2007 the government had reportedly set aside US$1.1 billion in subsidies to support the expansion of biofuel production infrastructure.[42] The government initiated a subsidized credit program for biofuel crop development[43] and projects that enhanced energy security,[44] along with income tax breaks on biofuel investments.[45] Biodiesel production capacity consequently approached 3 billion liters in 2009,[46] with 11 commercial scale biodiesel producers as of 2008.[47] The government has also instituted an export tax as well as a fertilizer subsidy.[48] In addition, the government

---

[38] 'Government Promotes CNG, Biofuel to Save Energy', Jakarta Post, May 21, 2006.

[39] Ibid.

[40] 'SE Asia Eyes Palm Oil', M&G Energy Watch, Dec. 19, 2006, http://news. monstersandcritics.com/energywatch/renewables/features/printer_1234800.php (accessed January 8, 2013) (also describes how government subsidizes the palm oil fuel blends to keep the industry competitive with diesel fuel).

[41] See GSI, supra note 34, at 17.

[42] Ibid. at 40.

[43] See Ministry of Finance Regulation No. 117/PMK.06/2006, 'Concerning Credit for Bioenergy Development and Estates Revitalization'. Under this program, the government, in cooperation with participating banks, would subsidize interest rates on loans made to qualifying plantation owners, effectively and significantly reducing the costs of borrowing. The subsidy for oil palm was 5%, which obviously made investments in the sector more attractive. See GSI, supra note 34, at 38–40.

[44] See Ministry of Finance Regulation 79/PMK.05/2007.

[45] GSI, supra note 34, at 40. (Examples of these incentives included a 30% reduction in tax liability from the total investment's income and accelerated depreciation and amortization options.) Ibid.

[46] Ibid. at 21.

[47] Ibid. at 22.

[48] See ibid. at 39; WBG Synthesis, supra note 29, at 13. Progressive export tariffs were instituted by the government to protect domestic biodiesel manufacturers, among others, when the global price of crude palm oil rose too high. As prices rose, specific taxes would be triggered to discourage exports. The government also subsidized fertilizers, so critical to the success of any plantation, to ease the financial burden on farmers and producers. These subsidies are reported to have reached US$1.65 billion in 2008. GSI, supra note 34, at 39.

continues to support a handful of state-owned enterprises, including Pertamina and PLN.[49] Both entities must consume excess biofuel production, and Pertamina must subsidize transport fuel and provide blended biofuels at the same price as (already subsidized) petroleum fuels.[50]

Complicating matters is the rising price of crude palm oil (CPO) in world markets, which is slowing investment in Indonesian biodiesel capacity.[51] Palm oil biodiesel has had to compete with conventional fossil fuel-based diesel, which garners its own generous subsidies from the government.[52] Indeed, it was partly the high cost of subsidizing petroleum, exacerbated by rising fuel prices and demand[53] that initially led to the government's interest in developing alternative forms of energy.[54] But high commodity and feedstock prices,[55] along with far higher costs of production, have slowed the transition to biofuels. Moreover, subsidies allocated for biofuel development have been far higher on paper than in actual disbursement.[56] While total budget commitments from 2006–08 were estimated at US$1.6 billion, actual support levels were most likely closer to US$197 million, suggesting serious logistical gaps.[57]

---

[49]   See GSI, supra note 34, at 18–21; see generally Business Watch Indonesia, 'Biofuel Industry in Indonesia: Some Critical Issues' (2007).

[50]   GSI, supra note 34, at 18–20.

[51]   See Ranjeetha Pakiam, 'Palm Oil Price May Rise to Highest Since 2008 as Output Slows, Ministry Says', Bloomberg, 13 Nov. 2011. Given the current high price of raw, unrefined palm oil, and the relatively low prices of petroleum, there is less of an incentive for biodiesel producers to assume the added costs of refinement when they could make a higher profit by simply selling the crude palm oil directly into the commodity and food markets. While global biofuel mandates and subsidies will keep demand for biofuels buoyant over the long term, rising food prices and increased demand (especially in developing countries) for crude palm oil will continue to divert resources in the short term from biodiesel investment. See Dewan, supra note 20.

[52]   Eric Unmacht, 'Faced with Soaring Oil Prices, Indonesia Turns to Biodiesel', Christian Science Monitor, 5 July 2006 (noting that Indonesia keeps the price of its petrol extremely low – at around 50 cents per liter for gasoline and diesel – through billions of dollars in subsidies). See also GSI, supra note 34, at 11 (the reduction in fuel subsidies in 2005 alone resulted in savings of US$10 billion).

[53]   Prices hit record highs of US$140 a barrel in July 2008. See US Energy Information Association, 'Crude Oil Spot Prices' 2008, eia.doe.gov (accessed January 8, 2013).

[54]   See, e.g., GSI, supra note 34, at 11.

[55]   Ibid. at 22 ('high feedstock prices have precipitated dramatic reductions in production levels, temporary suspension of operations and permanent cancellations or closures').

[56]   WBG Synthesis, supra note 29, at 13.

[57]   GSI, supra note 34, at 41.

D.   PUBLIC/PRIVATE FUNDING AND PARTICIPATION BY FUNDERS AND/OR INTERNATIONAL ORGANIZATIONS

As discussed above, the government, through direct subsidies, incentive programs, and Pertamina's biodiesel blending requirements,[58] incurred substantial costs in order to promote the biodiesel industry. At about US$200 million per year, the government unequivocally bears a far heavier burden than the private sector in supporting the industry.[59] And while the public (i.e. nationally owned) estates' share of crude palm oil production is decreasing, vertically integrated, private palm oil producers are taking a greater share of the market.[60] Private companies have now come to own 49 percent of Indonesia's oil palm plantations and have also invested in biofuel infrastructure capacity.[61] With biodiesel prices rising, the large palm oil producers are driving investment.[62] The industry itself is highly centralized, with only a few companies controlling a majority of the industry, many of which are still complicit in forced or inequitable labor treatment and deforestation.[63]

The Ministry of Forestry has had a lengthy record of favorable treatment towards private industry, understandable in light of the significant revenues provided by concessions. Even today, the Ministry's approach towards conservation is seemingly at odds with the President's endorsement of REDD+ projects,[64] which in turn has led to confusion in the

---

[58]   See Foreign Agriculture Service, US Dept. of Agriculture, 'Indonesia: Biofuels Annual, Annual Report 2009' 1 (2009). In 2008, the Ministry of Energy and Mineral Resources effectuated Regulation No. 32/2008, which mandated a minimum of 1% biofuel blending in fuel sold by the state in petrol stations throughout the country. The minimum blending percentages rise steadily through 2025 for both ethanol and biodiesel.

[59]   See GSI, supra note 34, at 41.

[60]   Chalil, supra note 31, at 8.

[61]   World Growth, supra note 1, at 11.

[62]   See, e.g., 'Corporate Power: The Palm-oil Biodiesel Nexus', Grain, 22 July 2007, http://www.grain.org/article/entries/611-corporate-power-the-palm-oil-biodiesel-nexus (accessed April 16, 2013); Credit Suisse, 'Biofuel Sector: Global Comparisons of a Fast-growing Sector', 30 Aug. 2006, http://www.ifqcbiofuels.org/pdfs/2006_09_Global_CreditSuisse.pdf (accessed January 8, 2013).

[63]   Melanie Pichler, 'Aseas, Agrofuels in Indonesia: Structures, Conflicts, Consequences, and the Role of the EU' 181 (2010).

[64]   See UN-REDD Programme, www.un-redd.org. REDD+ represents the UN Program 'Reducing Emissions from Deforestation and Forest Degradation in Developing Countries,' and focuses on the collaborative efforts and initiatives by

implementation of deforestation schemes.[65] The military has often been involved in the forced displacement of traditional landowners.[66]

A number of international organizations have played an increasingly important role in the sustainable development of the biodiesel industry. For instance, the World Wildlife Fund (WWF) has spearheaded the development of the Roundtable on Sustainable Palm Oil (RSPO) that encourages companies to use sustainably produced sources.[67] The World Resources Institute (WRI) has similarly facilitated sustainable management through Project POTICO[68] that aims to divert oil palm plantations to degraded lands and has developed advanced methodologies for inventorying such areas.[69]

Numerous organizations have analysed the social and environmental impact of palm oil development generally, including Greenpeace and the United Nations Environment Program.

In addition, several European Union countries have been working with Indonesia to expedite the transition to sustainably sourced palm oil products, including biodiesel.[70] The EU's Renewable Energy Directive of 2009 that requires 10 percent of transport energy to come from renewable sources includes biofuel sustainability criteria, specifically considering greenhouse gas emissions and whether sources are derived from lands of significant biodiversity value or high carbon stocks (e.g. continuously forested areas and peatlands).[71] And increasingly, individual states

---

participating countries to do so. See Section I(1)(j), infra, for more details on the REDD+ program.

[65] See, e.g., Marigold Norman, 'Rimba Raya Debacle Casts Pall Over Indonesian REDD', Forest Carbon Portal, 12 Sept. 2011, available at http://www.forestcarbonportal.com/content/rimba-raya-debacle-casts-pall-over-indonesian-redd (accessed January 8, 2013).

[66] See, e.g., S. Marti, Friends of the Earth, 'Losing Ground: The Human Rights Impacts of Oil Palm Plantation Expansion in Indonesia' 108 (2008).

[67] World Wildlife Fund – Promoting sustainable palm oil, http://wwf.panda.org/what_we_do/footprint/agriculture/palm_oil/solutions/ (accessed April 16, 2013). See also World Wildlife Fund – How clean is our palm oil?, http://www.wwf.org.uk/what_we_do/safeguarding_the_natural_world/forests/forest_conversion/how_clean_is_our_palm_oil_.cfm (accessed April 16, 2013).

[68] http://www.wri.org/project/potico (Project POTICO: 'Palm Oil, Timber and Carbon Offsets in Indonesia') (accessed April 16, 2013).

[69] http://www.wri.org/project/potico/faq (accessed April 16, 2013).

[70] See Section G, infra, for information on the major Norwegian initiative to promote biodiesel sustainability.

[71] EU Directive, supra note 16, art. 17. See also art. 18–19, http://ec.europa.eu/energy/renewables/biofuels/sustainability_criteria_en.htm (accessed April 16, 2013). Unfortunately, biofuels failing to meet the criteria may still be imported,

are beginning to commit to sourcing only certified sustainable palm oil (CSPO).[72] The Roundtable on Sustainable Palm Oil (RSPO) is one of the major players in this certification market, having instituted a set of Principles and Criteria used to ensure sustainability standards across the palm oil supply chain and independently certify participating stakeholders.[73] Such criteria include commitments to transparency, compliance with applicable laws and regulations, the use of appropriate best practices, environmental conservation and responsibility, fair labor and community standards, and the responsible development of new operations.[74] While palm oil biodiesel capacity is mainly financed by Indonesian, Malaysian, Chinese, and Singaporean investments,[75] the progressively advancing standards of buyers as large as the EU will hopefully continue to push Indonesian producers towards more sustainable models.

International financial institutions have also worked to support Indonesia's palm oil infrastructure. Traditionally, the World Bank Group (WBG) was highly supportive of efforts to stimulate sustainable economic development, including support for responsible worker transmigration initiatives.[76] But often loans and subsidies offered to oil palm companies were blind to the rampant occurrence of adverse social and environmental consequences. Recently, though, the WBG adopted a World Bank Group Framework and International Finance Corporation (IFC) Strategy to guide future engagement with and require sustainability within the palm oil sector.[77] This Framework 'will give priority to initiatives that encourage production on degraded lands and seek to improve productivity of

---

but will not be eligible for certain incentives under the program. See 'EU Directive Not Likely to Affect Malaysia's Palm Oil exports', ICIS, 24 May 2010, http://www. icis.com/Articles/2010/05/24/9361838/eu-directive-not-likely-to-affect-malaysias-palm-oil-exports.html (accessed April 16, 2013).

[72]   See, e.g., 'Manifesto of the Taskforce Sustainable Palm Oil', Nov. 2010, www.taskforceduurzamepalmolie.nl (accessed April 16, 2013).

[73]   See RSPO, 'RSPO Principles and Criteria for Sustainable Palm Oil Production: Guidance Document' (2006), http://www.rspo.org/files/resource_ centre/RSPO%20Criteria%20Final%20Guidance%20with%20NI%20Document. pdf (accessed April 16, 2013).

[74]   Ibid.

[75]   Pichler, supra note 63, at 181.

[76]   See Section E, infra.

[77]   See 'World Bank Group Adopts New Approach for Investment in Palm Oil Sector', 3 Apr. 2011, http://web.worldbank.org/WBSITE/EXTERNAL/TOPICS/ EXTARD/0,,print:Y~isCURL:Y~contentMDK:22882883~menuPK:336688~ pagePK:64020865~piPK:149114~theSitePK:336682,00.html (accessed April 16, 2013).

existing plantations.'[78] It also gives priority to those arrangements that benefit smallholders and rural communities, and aims to strengthen smallholder market access, contractual leverage, and skills capacity.[79] While some trade groups criticized the Framework for relying upon social and environmental safeguards to guide policy decisions,[80] the noted shift in policy is an encouraging sign of progress and a model for other organizations.

E. PERSONNEL TRAINING AND SOCIAL PROBLEMS

The cultivation of oil palm is heavily labor-intensive, and as such requires large numbers of unskilled agricultural workers.[81] In 2006, around 2 million people worked in the palm oil industry,[82] on plantations totaling 6.1 million hectares.[83] By 2008, over 41 percent of plantations were owned by small landholders ('smallholders'), with another 49 percent owned by large landholders and the remaining 10 percent by the government.[84] Yet smallholder areas have continued to grow rapidly – by nearly 12 percent per year from 1997 to 2007[85] – and are projected to increase by 64 percent between 2006 and 2015.[86] Unfortunately, a history of exploitation is deeply engrained within the fabric of current economic growth. Addressing long-standing gaps in both training and technical capacity for these smallholders is thus critical to the future of conservation efforts in Indonesia.

Under the direction of President Suharto, the government released tracts of valuable state land and offered financial subsidies (e.g. concessionary loan rates) to companies interested in developing oil palm plantations.[87] In

---

[78] Ibid.

[79] International Finance Corporation, http://www.ifc.org/ifcext/agriconsulta tion.nsf/Content/Home (accessed January 8, 2013).

[80] See Rhett Butler, 'Palm Oil Lobby Attacks World Bank's New Social and Environmental Safeguards', 18 Apr. 2011, www.mongabay.com (accessed January 8, 2013) (the trade groups asserted that the framework 'elevates radical ideological opposition to agriculture development above the needs of the poor and hungry in Africa').

[81] WBG Synthesis, supra note 29, at 29.

[82] World Growth, supra note 1, at 5.

[83] Indonesian Palm Oil Board, supra note 27, at 4.

[84] World Growth, supra note 1, at 11.

[85] WGB Synthesis, supra note 29, at 4.

[86] Ibid at 6.

[87] Ibid at 7–8.

return for their allocation, individual companies were required to employ the 'Nucleus Estate and Smallholder' (NES) model, through which the private developers ('nucleus') would plant oil palm plots on behalf of smallholders in that area.[88] Once the plantations matured, plots were then transferred to the smallholders for tending under the supervision of the nucleus developers.[89]

In order to ensure an adequate labor supply the government expanded its national transmigration scheme, whereby migrant workers from the more densely populated areas of Indonesia were relocated to plantations throughout the country,[90] but especially to the resource-rich and sparsely populated outer islands.[91] Eventually the success of this labor-transfer model led to formal institutionalization through a number of executive decrees,[92] and official endorsement from both the World Bank and the Asian Development Bank.[93]

The failures of this transmigration program under Suharto tie closely to those of the palm oil industry generally, including an embarrassing record of human rights violations and environmental devastation (as discussed at length below).[94] Oftentimes, 'indigenous peoples were inserted into the Transmigration schemes either by being resettled [to] Transmigrant villages made up of local people . . . or by being slipped into mixed settlements.'[95] But the failure to train or educate transmigrants throughout the course of the program was one of its greatest shortcomings, and led to many of the environmental harms perpetrated during its tenure. One author concluded that the program 'redistributed poverty, leaving the majority of transmigrants worse off due to totally inadequate planning and site preparation,

---

[88]   Van Gelder, supra note 21, at 19.
[89]   Van Gelder, supra note 21, at 19.
[90]   See WBG Synthesis, supra note 29, at 7–8; Van Gelder, supra note 21, at 19.
[91]   Indeed, plantations are concentrated in Sumatra, Kalimantan, Sulawesi, and Papua. See Colchester et al, supra note 11, at 24; M. Adriana Sri Adhiati, 'Down to Earth, Indonesia's Transmigration Program: An Update', 2001, http:// dte.gn.apc.org/ctrans.htm (January 8, 2013).
[92]   See, e.g., Presidential Decree 1/1986 (PIR-TRANS) (linked the NES model to the national transmigration program); Ministry of Agriculture Decree 853/1984 (private companies required to use NES model). See also WBG Synthesis, supra note 29, at 7–8.
[93]   See, e.g., Michael Goldman, 'Imperial Nature: The World Bank and Struggles for Social Justice in the Age of Globalization' 299 (2006); Adhiati, supra note 90.
[94]   See generally, 'Banking on Disaster: Indonesia's Transmigration Programme' 16(2/3) The Ecologist 57–117 (1986).
[95]   Colchester et al, supra note 11, at 44.

poor access to markets and neglect of soil and water properties indispensible for a prosperous agricultural economy.'[96] It was directly as a result of this failure to develop training capacity that Indonesia's natural resources were subjected to such shocking abuse. Transmigrants with no education or knowledge of the area, land and soil conditions, or best agricultural practices were left to work the land, with no appreciation for sustainable land management.

Over the next several years, the transmigration program began to contract while the oil palm industry was deregulated and increasingly privatized.[97] The fall of Suharto heralded an eventual shift to the decentralization of land use licensing rights to provincial governments.[98] This led to the current decentralized, pro-investor plantation model that allocates only 20 percent of concessions to smallholders and perpetuates the erosion of landowner rights.[99] Smallholders continue to lack the upfront capital, access to infrastructure (e.g. transportation and processing facilities), and training to successfully employ sustainable land conversions.[100] And while the mostly unskilled laborers are generally paid the provincial minimum wage,[101] payments are often insufficient to cover basic living expenses. Companies will frequently employ underpaid casual laborers to keep wage pressure on other employees.[102] Problems have also arisen with regard to the gross inequities in bargaining power that often undermine contract negotiations.[103]

F. PUBLIC PARTICIPATION PROVISIONS

A record of massive land rights abuses has marred the history of the oil palm industry, particularly as related to indigenous peoples and local communities.[104] In several case studies, 'no community members acknowledged being involved in the spatial planning exercises that allocate areas for the development of oil palm estates' on their lands, nor

---

[96]   Adhiati, supra note 90.
[97]   Colchester et al,. supra note 11, at 45.
[98]   Ibid.
[99]   Claude Fortin, 'The Biofuel Boom and Indonesia's Oil Palm Industry' Land Deals Politics Initiative Conference, 6–8 Apr. 2011, at 8.
[100] See, e.g., WBG Synthesis, supra note 29, at 33.
[101] Ibid. at 29.
[102] Ibid.
[103] Ibid. at 32.
[104] Colchester et al., supra note 11, at 173; Fortin, supra note 98, at 1.

were consensus-building meetings attempted.[105] Whether authorized at the national,[106] provincial, or local levels, government officials often permitted plantations to be developed in conscious disregard of preexisting ownership rights, and they sanctioned forcible and manipulative land acquisition.[107] Local governments usually have relied solely upon district spatial plans when making decisions, without any meaningful participation by locals or traditional communities.[108]

One of the reasons for this mistreatment is tied to the nature of indigenous customs – for instance, many practice transient cultivation rather than equally problematic fixed, plantation-based monoculture[109] Community leaders are often confused as to the nature and extent of land ownership transfers, which they understand to be temporary but in reality extinguish their rights.[110] Many 'public' or 'state' lands provide livelihoods and services for millions of rural Indonesians,[111] a fact that both government and private sector officials have freely ignored in negotiations with those rural citizens. Manipulation of the state's land use designation system, where tracts of lands are rezoned as 'state forest areas' to be slated for conversion, has led to flagrant bypassing of indigenous rights over land and resources.[112]

G.   ENVIRONMENTAL ASSESSMENT

The maturation and expansion of Indonesia's palm oil industry has come largely at the expense of the country's vast rainforests and peat bogs.

---

[105]   Colchester et al., supra note 11, at 170–171. (Unfortunately, Indonesian law does not mandate public participation provisions in this context. Moreover, the constitutional protections related to the land and customary rights of indigenous people are ambiguous and conflicting, which usually results in the abrogation of such rights.) See id., Chapter 3.

[106]   Under the Indonesian Constitution, the government has complete authority over land and natural resources, to be managed in the best interest of the public. See 'Indonesian Constitution of 1945', Article 33. See also Colchester et al., supra note 11, at 48–55, 170.

[107]   Colchester et al., supra note 11, at 174–175.

[108]   Ibid. at 170.

[109]   WBG Synthesis, supra note 29, at 30.

[110]   Colchester et al., supra note 11, at 16.

[111]   Fortin, supra note 98, at 4.

[112]   Colchester et al., supra note 11, at 177–78; 'Palm Oil Giant Misled the Public on Violent Conflict with Local Communities', Mongabay, 21 Nov. 2011, http://news.mongabay.com/2011/1120-fpp_asiatic_persada.html (accessed April 16, 2013).

The rapid growth of the biodiesel industry, in the absence of sustainable practices, conservation measures, and careful resource management, will surely intensify the already-dire environmental situation. Oil palm plantations have largely been centered on the island of Sumatra (constituting about 80% of total palm output), but are also found in Kalimantan, Papua, Sulawesi, and others.[113] The rainforests and peatlands native to these islands are some of the most important centers of biodiversity in the world, providing critical habitat and other ecosystem services for countless species.[114]

In the face of these vast natural resources, the pace of destruction has been truly staggering, with about 70 percent of oil palm plantations located on converted rainforest.[115] By some estimates, Indonesia suffered losses of around 28 million forest hectares from 1990 to 2005.[116] Today, two-thirds of oil palm expansion is based on the conversion of virgin rainforests, and a quarter of that area is on soil created by the destruction of peat bogs.[117] Peat soil that is compressed, drained, and burned for agricultural purposes causes irreparable damage and leads to enormous emissions of greenhouse gases (due to the highly stored carbon content in the peat bogs).[118] Despite the fact that development costs are 30 percent higher on peat lands than on mineral soils,[119] peat forests in Sumatra, Papua, and Kalimantan are cleared at rates of almost 100,000 hectares per year.[120] Meeting projected growth figures for palm oil biodiesel would require three times the area currently planted, totaling more than 15 million hectares of mature plantations.[121]

Ecologically, the loss of forest cover has resulted in a litany of adverse biological crises with cascading effects.[122] Plantation conversion 'eliminates

---

[113] WBG Synthesis, supra note 29, at 11; Pichler, supra note 63, at 179.

[114] See, e.g., Zoological Society of London, 'The Conservation of Tigers and Other Wildlife in Oil Palm Plantations' 11 (2007) (hereinafter ZSL) (noting that Indonesia accounts for 17% of the planet's birds, 12% of its mammals, and 10% of its flowering plants).

[115] World Bank Group, supra note 8, at 17.

[116] Rhett Butler, 'Indonesia's plan to save its rainforests', Mongabay, 14 June 2010, http://news.mongabay.com/2010/0614-indonesia_purnomo_saloh.html (accessed April 16, 2013).

[117] UNEP, supra note 18, at 65. (Some have forecast that land from digging up peat bogs will comprise a 50% share by 2030.)

[118] See WBG Synthesis, supra note 29, at 21–22 (peat soils cultivated with oil palm emit around 73 tons of carbon per hectare per year).

[119] Ibid. at 19.

[120] Ibid.

[121] Greenpeace, 'How the Palm Oil Industry is Cooking the Climate' 8 (2007).

[122] See, e.g., WBG Synthesis, supra note 29, at 17.

entire niches in the forest, destroying the canopy, annihilating complex nutrient cycling systems, and turning clear-flowing streams into squalid repositories of fertilizers, pesticides, and industrial chemicals.'[123] Monoculture requires more fertilizers and pesticides, which inevitably end up in water supplies as effluent and runoff.[124] Deforestation and peatland destruction result in severe degradation of watersheds, compromised water quality, and soil erosion and degradation.[125] This loss of water retention capacity results in flooding, landslides, and the drying associated with lower dry season base flows.[126] And with significant drying comes the heightened risk of uncontrolled fires.[127] Contributing to the severity of this problem is the illegal but oft-employed practice of clearing rainforests through the use of fire.[128] These fires, both wild and intentional, drive the perpetuation of forest degradation cycles and exacerbate greenhouse gas (GHG) emissions.

The government estimates that the monumental losses in terms of deforestation, forest degradation, and the burning of peat lands constitutes about 80 percent of national emissions.[129] Consequently, Indonesia's GHG emissions have skyrocketed to the third greatest in the world.[130] One study estimated that draining peatland in Indonesia accounts for 660 million tons of carbon per year, with fires contributing another 1.5 billion tons annually.[131] Researchers have found that when the land upon which

---

[123]   Rhett Butler, 'Greening the World with Palm Oil?', Mongabay, 26 Jan. 2011, http://news.mongabay.com/2011/0126-palm_oil.html (accessed April 16, 2013).
[124]   GSI, supra note 34, at 52. See generally, Van Gelder, supra note 21.
[125]   WBG Synthesis, supra note 29, at 17; Blue Green Alliance, 'Illegal Logging in Indonesia: The Environmental, Economic and Social Costs' 5–6 (2010).
[126]   WBG Synthesis, supra note 29, at 19.
[127]   Blue Green Alliance, supra note 124, at 5–6.
[128]   See, e.g., ZSL, supra note 113, at 14 (in 1997, the government accused 133 oil palm companies of burning about 12 million hectares of rainforest); WBG Synthesis, supra note 29, at 19–20 (noting that 'smallholders continue to use fire to clear land because they lack the capital needed to pay for heavy land clearing machinery').
[129]   Ministry of Forestry, Republic of Indonesia, 'Reducing Emissions from Deforestation and Forest Degradation in Indonesia – REDD Methodology and Strategies Summary for Policy Makers' (2007).
[130]   See World Bank, 'Indonesia & Climate Change' (June 2007). See also Oxfam, Oxfam Briefing Paper, 'Another Inconvenient Truth: How Biofuel Policies are Deepening Poverty and Accelerating Climate Change' (2008) (finding that by 2020, the combined emissions of palm oil production in Indonesia and Malaysia, processed only to satisfy EU biofuel demands, would equate to around 4.6 billion tons of CO_2).
[131]   Elisabeth Rosenthal, 'Once a Dream Fuel, Palm Oil May Be an Eco-Nightmare', NY Times, 31 Jan. 2007.

oil palm for biodiesel has been converted from rainforest, the resulting greenhouse gas emissions are 800 percent higher as compared to conventional biodiesel; that figure increases to 2,000 percent if the land is changed from peat forest.[132] Lifecycle analyses have demonstrated that biodiesel produced from palm oil grown on peatlands may take more than 400 years to produce any emissions savings when compared to fossil fuels.[133] Given the fact that oil palms produce over about 25 years,[134] such time frames would seem to reflect grossly overestimated carbon reduction benefits.

Rainforest conversion invariably leads to greater accessibility to the area, and a corresponding increase in poaching, development, pollution, and human settlement.[135] This caustic fragmentation of critical habitat in areas of such abundant biodiversity has inevitably resulted in conflict with – and the subsequent decline of – dozens of species.[136] One of the highest profile conflicts involves the orangutan,[137] as over half of the world's wild population has disappeared in the last 20 years alone.[138] Other large species – including most notably rhinos, elephants, and Sumatran tigers – have suffered critical drops in population count and habitat availability.[139] For instance, Sumatra's Riau province has lost 65 percent of its forests

---

[132] UNEP, supra note 18, at 53. See also, Intl. Council on Clean Transportation, 'Review of Peat Surface Greenhouse Gas Emissions from Oil Palm Plantations in Southeast Asia' 3 (2011) (finding that palm oil produced on peatland releases 86 megatons of CO_2 per hectare per year over a 50-year period).

[133] Butler, supra note 122.

[134] UNEP, supra note 18, at 54.

[135] WBG Synthesis, supra note 29, at 18; ZSL, supra note 113, at 14. (Once rainforests have been converted, the resultant cropland is understandably poorly suited for habitat, with strong oil palm intolerance observed in the most endangered of species.) See ZSL, supra note 113, at 43. According to this study, only 10% of mammals recorded in the plantation showed any ability to survive on a long-term basis.

[136] See WBG Synthesis, supra note 29, at 18 (noting that large scale, monoculture plantations ranging in size from 4,000 to 20,000 hectares can result in 'serious loss of connectivity between any remaining forest patches').

[137] See generally, UNEP, 'The Last Stand of the Orangutan: State of Emergency' (2007).

[138] Ibid. Over 80% of orangutan habitat has already been destroyed or depopulated. Birute Mary Galdikas, 'The Vanishing Man of the Forest', NY Times, 6 Jan. 2007, http://www.nytimes.com/2007/01/06/opinion/06galdikas. html?pagewanted=1 (accessed April 16, 2013). See also Butler, supra note 122 (placing orangutan death estimates at 1,500 to 5,000 per year, out of 54,000 living in Borneo and only 6,500 in Sumatra. Orangutan habitat in Kalimantan has fallen from 55,000 square miles in 1992 to fewer than 27,000 square miles today. Similarly, Sumatra has lost more than 90% of its primary forest cover since 1975).

[139] See generally, Greenpeace, supra note 120.

over the last 25 years, resulting in declines of elephant populations of 84 percent (to 210 individuals) and tiger populations by 70 percent (to 192 individuals).[140]

H.   NET SAVINGS AND COSTS

As has been stated, total reported biofuel budget allocations from 2006 through June 2008 were estimated at US$1.6 billion.[141] Total actual support over the same time period equaled about US$200 million.[142] Pertamina, the state-owned fuel enterprise, was mandated to sell biofuel blends at the same price as subsidized petroleum fuels.[143] Due to these requirements, the enterprise announced losses over 2006–08 totaling US$40 million.[144] Because of the consistently high cost of biodiesel production over the last several years, the industry has been increasingly unprofitable. And biodiesel development is expected to require additional investments of US$2.3 billion through 2025.[145] The subsidies for biodiesel will be raised in 2012 from 2,000 Indonesian rupiah (IDR) per liter to 2,500–3,000 IDR per liter, and Pertamina is expected to allocate 58.5 billion IDR (US$6.6 million) to establish biofuel infrastructure throughout the country.[146]

Environmental services are frequently marginalized in favor of more immediate economic development benefits. Costs, in addition to environmental externalities, include the opportunity costs of investing in biodiesel instead of some other source of renewable fuel. Further, biodiesel production often competes with food producers over access to arable land, and can consequently exert upward pressure on food prices[147] in a country with chronic food shortages.[148] This is a serious

---

[140]   Blue Green Alliance, supra note 124, at 5–6.
[141]   GSI, supra note 34, at 41.
[142]   Ibid. at 60.
[143]   See ibid. at 37.
[144]   Ibid. During this time period, the same subsidies were offered to both fossil fuels and biofuels. Because Pertamina was required to sell the biofuels at the same price as petroleum fuels, massive losses were sustained each time biofuels were more expensive to procure than petro fuels (as they often were during this period). See also Foreign Agricultural Service, supra note 10.
[145]   Anasia Silviati, US Commercial Serv., 'Indonesia: Biofuel Development' 1 (2008).
[146]   Foreign Agricultural Service, US Dept. of Agriculture, 'Indonesia: Biofuel Annual' 2 (2011).
[147]   Thoenes, supra note 9, at 7; GSI, supra note 34, at 46.
[148]   See Food and Agriculture Organization, 'Countries in Crisis Requiring

problem and may eventually lead to political instability and greater aid needs.

I. PROBLEMS AND SOLUTIONS

1. Critical Issues

a. Inadequate consideration of environmental costs

Perhaps the singular defining feature of Indonesia's palm oil program (including biodiesel production) has been its disastrous impact on the country's pristine rainforests and peatlands. It is undeniable that environmental concerns were almost entirely neglected over the duration of the palm oil program. While the international community has helped to focus more attention upon these issues, more concerted action must be taken in order to keep pace with projected demand for biodiesel worldwide. Moreover, establishing pilot projects and examples of sustainable management in a country with significant preexisting development can serve to illustrate successes and failures for other emerging economies considering oil palm for biofuel.

Ensuring that new, credible, and powerful safeguards are in place at the governmental level is crucial to the process of internalizing environmental damage. The relatively new Environmental Management and Protection Law is a step in the right direction.[149] Among other provisions, the law calls for the integration of environmental concerns into the spatial planning process;[150] requires environmental permitting and risk analysis to supplement the preexisting but often ill-enforced environmental impact assessment, along with the setting aside of security funds in the case of environmental damages; and introduces both harsher penalties and enhanced legal remedies that can be enforced by local communities and environmental NGOs.[151] But strong government enforcement

---

External Assistance', www.fao.org/docrep/010/ah881e/ah881e02.htm (accessed April 16, 2013).

[149] No. 38/2009 Environmental Management and Protection Law.

[150] See WBG Synthesis, supra note 29, at 23.

[151] See Soemadipradj and Taher, 'New Environmental Law – Better Protection or More Legal Hurdles for Industry', Asia Legal Business: Legal News (2010). (The Environment Ministry has vowed much better enforcement, and is working with the courts and its own civil servants to increase capacity and training.) See Fidelis Satriastanti, 'Ministry Vows Better Enforcement as 2009 Environmental Law Takes Hold', Jakarta Globe, 4 Oct. 2011.

and implementing regulations also will be needed to support this law.[152] Efforts must also be made to avoid converting areas with high environmental or conservation value. The High Conservation Value Forests (HCVF) concept, first adopted by the Forest Stewardship Council, should be relied upon to inform future decision-making with regard to industry expansion.[153] Ascertaining which lands are of such high ecological value will necessitate more and improved scientific research, which should be supported through government and international funding.

Finally, the government must invest in continuing education programs that can disseminate information to smallholders and other stakeholders on services provided by the natural environment, zero-burning practices for replanting, and conservation tactics that protect biodiversity and minimize the impact upon critical ecosystems.[154] Identifying and quantifying the value of ecosystem services to communities will help to make the case for valuable conservation initiatives and REDD-type projects. Ecosystem restoration concessions, through which land permittees are authorized to sell ecosystem services in exchange for sustainable management of the land or restoration projects, are one of the ways to monetize the value provided by natural services.[155] Improving the technical capacity to manage these services, including most importantly water resources, is necessary to ensure proper protection by provincial governors and smallholders alike. Financial incentives for avoiding destruction of peatlands and rainforests and establishing conservation set-asides to protect wildlife corridors should also be employed.

---

[152] The implementing regulations have only recently been passed in October 2011.

[153] See WBG Synthesis, supra note 29, at 25. (The High Conservation Value Forests (HCVF) concept is used to describe and identify those forests with extremely high ecological value deserving of conservation prioritization. There are many criteria used to assess such forests, including the level of biodiversity natural to the area, the presence of threatened and endangered species, and the existence of natural resources (e.g. water filtration) critical to the health of the ecosystem and local communities).

[154] American Palm Oil Council, 'Sustainable Practices' (2008), www.americanpalmoil.com/sustainable-pome.html (accessed April 16, 2013); GSI, supra note 34, at 56–57.

[155] Jeremy Hance and Rhett Butler, 'Converting Palm Oil Companies from Forest Destroyers into Forest Protectors', Mongabay, 2 Jan. 2011, http://news.mongabay.com/2011/0103-wri_interview_hance_butler.html (accessed April 16, 2013).

b.   Use of degraded land

Degraded land refers to those areas that have been stripped of natural vegetation cover, losing productive capacity as a result of land use changes that include logging and cultivation.[156] It is estimated that of the 71 million hectares of degraded land in Indonesia, 30 percent has been induced through human agricultural activities.[157] It is essentially because such lands are lower in productive capacity (for agriculture or habitation) that they serve as attractive alternatives to development in virgin forests or peatlands. Likewise, developing already degraded lands is less capital-intensive and thus makes a more attractive investment. Careful management of degraded lands does have the potential to support projected industry expansion by 2020.[158]

But the government's treatment of such degraded lands has often been problematic. Most glaringly, the government has never established a complete, transparent inventory of degraded lands throughout the country. Still, the Department of Agriculture has argued that there are about 27 million hectares of 'unproductive forestlands' that may be available for conversion.[159] But in the absence of a national inventory of lands, many lands are too often misclassified and undervalued. This leads to chronic marginalization of lands with underestimated value, in terms of both production and biodiversity.

WRI has been at the forefront of efforts to expand oil palm plantations onto degraded lands and grasslands.[160] Through its POTICO program,[161] WRI has encouraged the shift towards degraded land development, aided in the development of mapping degraded lands, and advised the industry how best to achieve sustainability and positive social outcomes.[162] Determining levels of degradation in relation to a mapping program is critical for the success of the overall effort, as some lands are far more seriously (and irrevocably) degraded than others.[163] Research has shown that even highly degraded forests retain more biodiversity than monoculture plantations, representing great value for conservation efforts.[164]

---

[156]   See World Growth, supra note 1, at19; Hance and Butler, supra note 154.
[157]   World Growth, supra note 1, at 19.
[158]   WBG Synthesis, supra note 29, at 18.
[159]   Colchester et al., supra note 11, at 25.
[160]   Hance and Butler, supra note 154.
[161]   http://www.wri.org/project/potico/faq (accessed April 16, 2013).
[162]   Hance and Butler, supra note 154.
[163]   Ibid.
[164]   WBG Synthesis, supra note 29, at 18.

c. Failure to properly inventory lands and associated titles

The government's failure to gather adequate and comprehensive information on public lands has led to a system exploitative of traditional land rights. Some estimates place the legal status of almost 90 percent of forests as unclear.[165] Less than 40 percent of rural lands in Indonesia are titled, 'a proportion which is declining year on year as new holdings are created faster than the national land office can survey and register them.'[166] By actively ignoring true ownership claims, the government has been continually complicit in the abuse of landowner rights by both state- and privately owned corporations.

The government has further failed to catalog and designate lands according to biological and productive capacity, which includes the identification of degraded lands. Funds have already been allocated to these projects, but progress has been slower than expected. Properly cataloguing lands in terms of ownership (traditional, customary, or otherwise), topography, and government jurisdiction must be completed quickly in order to make progress on other substantive fronts. Furthermore, the government must invest in a workable system of titling that respects and legally acknowledges traditional, customary land usages and communal ownership schemes.

d. Lack of effective oversight and monitoring

The environmental and economic uncertainties associated with the production of palm oil biodiesel calls for the institution of a strong oversight mechanism for calculating lifecycle greenhouse gas assessments, documenting sustainable production processes, and verifying the success of forest protection projects. It is only once initial baselines are established that true progress may be measured and analysed. Such independent monitoring, reporting, and verification (MRV) mechanisms are central to the funding of GHG reduction strategies, in order to reassure investors of accountability and transparency. Considering the debatable emissions benefits of biodiesel and its potential to conflict with food security, objective MRV systems must be instituted so that policymakers have a clear picture of the true costs and benefits associated with biodiesel production programs on a national scale.

The government must tighten oversight mechanisms related to the implementation of national regulations and directives by provincial and

---

[165]   Van Gelder, supra note 21, at 65.
[166]   Marcus Colchester, 'Indonesia: Putting Rights into Forest Conservation', Arborvitae, 36 (2008).

local leaders. Worryingly, many of the issues discussed above signal a striking breakdown in the rule of law. For instance, lands encumbered with ownership rights have been illegally allocated to third parties for conversion to palm oil. In a disturbing but common trend noted above, concessionaires harvest valuable timber supplies from the area but fail to then develop the more costly oil palm plantation at all.[167] Sixty percent of recorded timber has been harvested illegally,[168] and only six of the 20 million hectares cleared in the name of oil palm have actually been planted. [169] Finally, illegal logging was found to take place in 37 of 41 Indonesian national parks. Reversing these trends through strong federal oversight and the implementation of advanced monitoring technology will determine the extent to which countries and industries are willing to invest in sustainability measures.[170]

e. Government mismanagement and ineffective coordination

Decentralization of the oil palm industry resulted in granting licensing authority to the provincial governors. As such, the federal government has often been unable to align regional political and economic interests with national environmental initiatives. The decentralization of authority over forest management activities has resulted in confusion, conflict between the institutional mandates at different levels of government, and an utter lack of accountability in land use decision-making.[171] Oftentimes, land is conceded at the district level without the knowledge or approval of pro-vincial or central governments. These unauthorized land grants have often resulted in illegal conversion and fraudulent land transfers.[172]

The national government must assert control over the management of forest resources. This will require more effectively directing the Ministry of Forestry, which continues to employ policies that conflict with national directives. Allegedly, the Ministry is 'highly skeptical' about the market for REDD-carbon credits.[173] But weaning the Ministry 'from its long-standing reliance on concessions to clear and develop many of the forests

---

[167]   See, e.g., Van Gelder, supra note 21, at 25.

[168]   WBG Synthesis, supra note 29, at 37. (Illegal logging has also been found to take place in 37 out of 41 Indonesian national parks surveyed.) See UNEP, supra note 136.

[169]   Van Gelder, supra note 21, at 67.

[170]   For instance, see Section D, on the EU's sustainability criteria regarding the import of biofuels.

[171]   World Growth, supra note 1, at 20.

[172]   Ibid.

[173]   David Fogarty, 'How Indonesia Crippled Its Own Climate Change Project', Reuters, 16 Aug 2011.

most valued for conservation and REDD projects will likely require more than just the flick of a pen.'[174] The federal government must be willing to exert authority and spend political capital to bring the Ministry into conformance with national policy goals and minimize questionable land transfers.

Studies have indicated that government willingness at all levels to make land available for conversion to oil palm has had a greater effect than credit and subsidy programs.[175] Because local governments have a strong revenue interest in promoting industry development, the system has often resulted in illegal deforestation within protected areas, fraudulent permitting, and widespread corruption. Indonesia Corruption Watch found that illegal permitting totals about US$2.3 billion per year.[176] Transparency within the industry also suffers, in light of the many opportunities for corruption and graft. Industry expansion and sustainable competition alike are thus affected.

Ensuring accountability at every level is crucial for restoring credibility to the resource management programs in question. The government must work with international watchdog organizations to implement effective regulatory reforms and reign in abusive practices. Wholesale reforms at the government level must strengthen smallholder capacity building and the overall efficacy of sustainable initiatives.

f. Corporate abuses

Due to the high capital and infrastructure costs of palm oil refinement and historically beneficial treatment at the hands of the government, large companies have had the opportunity to dominate the market at the expense of individual farmers. Today, private industry is a powerful vested interest within the sector. This setup has created grave imbalances in negotiating power; stoked countless unresolved conflicts and disputed territory claims; and amassed a record of flagrant human rights and environmental abuses. Such companies have often relied 'on deception, coercion, and violence to quell opposition and to allow for continued expansion at an unbridled pace.'[177] Regrettably, these practices have often been employed with the blessings (and sometimes aid) of the government.

The need to radically reform this model is indisputable. Indeed, tackling corruption at the highest levels of government presents a great challenge,

---

[174] Norman, supra note 64.
[175] WBG Synthesis, supra note 29, at 13.
[176] Fogarty, supra note 172.
[177] Fortin, supra note 98, at 1.

given the scale of industrial plantations and the expected pace of future development. Palm oil companies have had several decades to entrench themselves within the political system, and will leverage the national desperation for both economic and energy security. Strict oversight and regulation of the industry will be necessary to ensure a more equitable distribution of both the costs and benefits of the program. The government must also be prepared to enforce environmental and labor laws more strictly, which will most certainly require additions to its enforcement capacity.

Moreover, pricing carbon effectively, either through the imposition of a carbon tax or through a cap and trade offsetting system, would force companies to internalize the costs of their enterprises. Supporting only organizations that promote conservation, sustainability, and best practices would help to drive the market towards greater accountability and transparency. The government must also work with the EU, US and emerging economies like China and India, all massive buyers of palm oil, to develop sustainability standards and enforce agreements. The purchaser countries could assert great leverage by restricting their purchases only to products that are certified as sustainable.

g. Capacity building and accessibility

Oil palm production can certainly yield positive socio-economic benefits in the form of expanded infrastructure, access to training, employment opportunities, tax revenues, and biodiesel for domestic consumption.[178] However, more often than not smallholders struggle to realize the true potential of oil palm initiatives. Smallholder plantations have 'significant lower average yields (hence lower financial returns) than estate- or government-owned plantations.'[179] The government, together with international partners, must continue to fund education and training programs for smallholders to address this imbalance. For example, the government has worked with the support of the World Bank to create the Farmer Empowerment through Agricultural Technology and Information (FEATI) Project, which empowers farmers 'through improved information networks, community agribusiness development, and enhanced linkages between research and extension.'[180] In coordination with such capacity building activities, the government should also invest in tax

---

[178] See, e.g., WBG Synthesis, supra note 29, at 27; World Bank Group, Working Paper: 'Poverty, Income, Inequality and Oil Palm Activity: A Technical Summary of District-level Empirical Analysis' 6 (preliminary draft) (2009).

[179] WBG Synthesis, supra note 29, at 34.

[180] WBG Synthesis, supra note 29, at 37.

incentives for those companies procuring sustainable palm oil supplies from rural farmers – similar to Brazil's Social Fuel Seal Program.[181] Under this program, biodiesel producers receive tax exemptions for sourcing a certain percentage of raw materials from smallholders, and in return farmers receive technical training, contractual rights, and access to subsidized credit.[182]

Ensuring familiarity with the concepts of conservation of both land and wildlife is the first step towards guaranteeing positive outcomes. Demonstrating the viability of revenue derived from such programs, though, is critically important. The government must also disseminate information related to best management practices, new agricultural inputs technologies, and improved seed varieties. Finally, ensuring access to accurate, clear information is necessary in order to generate public buy-in and accountability. Involvement in public decision-making and spatial planning processes tend to reinforce communal roles and strengthen public support. By strengthening the connection between the land and smallholders, a long-term interest in maintaining and protecting its economic and natural viability is cultivated. Indeed, given the rapid rise in smallholder ownership – which has come to constitute a majority – emphasizing smallholder education is an investment in the future of sustainability. Opening up the planning process to public participation will serve to check the erosion of socio-economic values by corporations and give greater weight to the autonomy of the Indonesian people.

h. Recent developments
Several recent developments in the international arena and across the palm oil sector lend hope and credibility to the government's progressive agenda. Many of these initiatives, projects, and agreements address some of the more pressing concerns identified above, and offer a useful framework for considering the future of palm oil investments.

i. Roundtable on sustainable palm oil
In response to local negative public reaction against the destructive practices of the palm oil industry, several of the major palm oil growers, processers, and refiners, along with conservation groups, financial

---

[181] See, e.g., Bioenergy & Food Security, Food and Agriculture Organization of the United Nations, 'Social Fuel Seal: Brazil' (2010); Biopact, An In-depth Look at Brazil's 'Social Fuel Seal,' (2007/3/23), http://news.mongabay.com/bioen ergy/2007/03/in-depth-look-at-brazils-social-fuel.html (accessed April 16, 2013).

[182] Ibid. (While the program is not perfect in practice, it can serve as a good model for others).

institutions, and other NGOs, created the Roundtable on Sustainable Palm Oil (RSPO).[183] The purpose of the RSPO is to reform palm oil production along more sustainable lines by using market mechanisms and initiatives.[184] The group also aims to increase transparency in the sector as well as protections for landowners; adopt and enforce labor, human, and land rights, both domestic and international; and minimize the environmental impacts of the industry on rainforests, peat bogs, and indigenous peoples.[185] The RSPO currently includes 40 percent of global palm oil production,[186] and has recently certified more than 25,000 palm oil smallholders as sustainable producers.[187] Sales of certified sustainable palm oil (CSPO) totaled 1.3 million metric tons in 2010.[188]

Despite marked progress, the RSPO has already encountered numerous problems with its suppliers illegally clearing rainforests, including some of the largest multinational companies in the industry.[189] Criticisms of the RSPO include lack of sufficient oversight and punitive capacity, conflicts of law, and the feeling by some that its true intent is to legitimize continued expansion of the industry.[190] Many feel that without significant funding or investigative powers, members have little actual ability to influence the destructive practices of their subsidiaries.[191] For instance, a subsidiary of RSPO-member Cargill, CTP Holdings, was found to have been operating two completely undisclosed plantations, both of which, according to the Rainforest Action Network (RAN), lack business, location, and timber cutting permits.[192] 'Greenwashing' seems to be a valid concern of both critics and supporters.[193]

---

[183] See www.rspo.eu (accessed April 16, 2013).
[184] See ibid.; WBG Synthesis, supra note 29, at 32.
[185] See www.rspo.eu (accessed April 16, 2013).
[186] See IFC, 'Market Transformation Strategy for Palm Oil' 4 (2008).
[187] 'RSPO Celebrates First 25,000 Certified Palm Oil Family Farms', 11 Nov. 2010, http://www.rspo.eu/docs/101111_rspo_pr_sh.pdf (accessed April 16, 2013).
[188] 'Sales of RSPO-certified Palm Oil Surge 225%', Mongabay, 10 Jan. 2011, http://news.mongabay.com/2011/0109-rspo_sales.html (accessed April 12, 2013) (this figure rose from 400,000 tons in 2009).
[189] See, e.g., Butler, supra note 122.
[190] Ibid.
[191] Pearce, supra note 5.
[192] See Jeremy Hance, 'Activists Lock Themselves in Cargill Headquarters as New Report Alleges Illegal Deforestation', Mongabay, 5 May 2010, http://news.mongabay.com/2010/0505-hance_cargill.html (accessed April 16, 2013).
[193] See, e.g., Food Climate Research Network, 'Palm Oil Company Commits to Halting Deforestation', http://www.fcrn.org.uk/research-library/industry-actions/retailer-announcements/palm-oil-company-commits-halting-deforestation (accessed April 16, 2013) (while Golden Agri-Resources Limited, a palm

Nonetheless, the RSPO has been addressing many of the issues discussed above through the voluntary Principles and Criteria, including the
adoption of the High Conservation Value Forests (HCVF) concept and
associated member requirements.[194] In 2010 the Netherlands (Europe's
largest palm oil trader) committed to source only CSPO-certified palm oil
by 2015, becoming the first nation to do so, but since joined by corporations such as Unilever, Wal-Mart, and others.[195] If influential companies
from the largest emerging economies – including China and India, where
demand for palm oil is expected to grow to exponentially[196] – begin to
purchase only CSPO-certified products, then the RSPO may yet help to
drive the global market towards true sustainability.[197]

j.   REDD+ and other greenhouse gas reduction strategies

Indonesia has been working with the international community to develop
a deal that will effectively address climate change.[198] One of the most critical areas of these negotiations – especially in light of increasing interest in
palm oil biodiesel – is the UN's forestry initiative, the Reducing Emissions
from Deforestation and Forest Degradation (REDD+) program.[199]

---

oil supplier, pledged to halt deforestation in high conservation value lands,
Greenpeace points out that the company's pulp and paper arm, Asia Pulp and
Paper, is still implicated in deforestation schemes).

[194]   WBG Synthesis, supra note 29, at 25. See note 152, supra, for more information on the HCVF concept; and Section D, supra, for a brief description of
RSPO sustainability criteria.

[195]   Matthew McDermott, 'Netherlands Will Only Use Certified Sustainable
Palm Oil by 2015 – First National-Level Commitment', Treehugger, 5 Nov. 2010,
http://www.treehugger.com/files/2010/11/netherlands-only-use-sustainable-palm-
oil-2015.php (accessed April 16, 2013); 'Sales of RSPO-certified Palm Oil Surge
225%', Mongabay, 10 Jan. 2011. The initiative was organized by the Dutch Task
Force Sustainable Palm Oil. For information on this program, see Task Force
Duurzame Palmolie, http://www.taskforceduurzamepalmolie.nl/ (accessed April
12, 2013).

[196]   Colchester et al., supra note 11, at 20.

[197]   Global demand for palm oil is set to double by 2020, mainly as a result of
increased interest from developing countries. See ibid. at 20.

[198]   See Rachmat Witoelar, Executive Chair, National Council on Climate
Change, 'Indonesia Voluntary Mitigation Actions', 30 Jan. 2010, http://unfccc.int/
files/meetings/cop_15/copenhagen_accord/application/pdf/indonesiacphaccord_
app2.pdf (accessed April 16, 2013) (target of 26% emissions reduction by 2020).
See generally US Agency for International Development, 'Indonesia Country
Report from Ideas to Action: Clean Energy Solutions for Asia to Address Climate
Change', (2007).

[199]   http://www.un-redd.org/ (accessed April 12, 2013). See also http://www.
un-redd.org/AboutREDD/tabid/582/Default.aspx (accessed April 12, 2013)

Current rates of deforestation worldwide are about 13 million hectares per year,[200] which if continued, will result in 286 million hectares deforested by 2030.[201] According to some projections, a land demand equivalent to 40–180 percent of ongoing deforestation would be required to meet future biofuel demand.[202] With rates of virgin rainforest destruction in Indonesia so high,[203] deforestation reduction mechanisms are critical to the viability of natural ecosystems moving forward. Through REDD+, developing countries can earn carbon credits in exchange for terminating forest and peat conversion, and thus preserving its carbon-storing capacity.[204]

A focus on emission mitigation options further supports the effectiveness of REDD+ schemes by enhancing the value of carbon reductions. Understanding this concept, President Yudhoyono announced in 2009 his country's intention to voluntarily cut GHG emissions by 26 percent by 2020 and up to 41 percent with support from developed countries.[205] The government has already mapped out an official Climate Change Action Plan[206] and is drafting a REDD+ National Strategy.[207] REDD+ programs have been funded under the Forest Carbon Partnership Facility,

---

(noting that REDD+ 'includes the role of conservation, sustainable management of forests and enhancement of forest carbon stocks').

[200] See Food and Agriculture Organization, 'Global Forest Resource Assessment 2010' (2010), http://www.fao.org/forestry/fra/fra2010/en/ (accessed April 16, 2013).

[201] UNEP, supra note 18, at 67.

[202] Ibid.

[203] David Smith, 'Five Years to Save the Orangutan', The Guardian, 24 Mar. 2007, http://www.guardian.co.uk/environment/2007/mar/25/conservation.theobserver (accessed April 16, 2013) (up to 98% may be destroyed by 2022).

[204] Tonka Dobreva and John Alexander Adam, 'Is There a REDD Solution to the Palm Oil Problem?', http://ezinearticles.com/?Is-There-a-REDD-Solution-to-the-Palm-Oil-Problem?&id=6503761 (accessed April 12, 2013) (the program was officially accepted by the international community at the United Nations Framework Convention on Climate Change Cancun Climate Summit, with US$30 million pledged to facilitate appropriate projects).

[205] Rhett Butler, 'Will Indonesia's Big REDD Rainforest Deal Work?', Mongabay, 28 Dec. 2010, http://news.mongabay.com/2010/1228-indonesia_redd.html (accessed April 16, 2013) (the 1.2 gigaton cut represents 8% of the total UN emissions target for 2020).

[206] See Yulia Suryanti, State Ministry of Environment, 'Indonesia's National Climate Change Action Plan and MRV', 18th Asia Pacific Seminar, 2–3 Mar. 2009; see also National Council on Climate Change, Presidential Regulation No. 16/2008, On the National Council on Climate Change (2008).

[207] See Indonesia's National REDD Strategy, UN-Redd Programme Newsletter, Sept. 2010, www.un-redd.org/Newsletter 12/Indonesia_REDD_Strategy/tabid/5533/Default.aspx (accessed April 16, 2013).

the Global Environment Facility, and other organizations.[208] The government has developed a Climate Change Reform Program,[209] one supported by the World Bank, the Asian Development Bank, and other financial institutions.[210] The Climate Investment Funds, a multi-donor mechanism supported by numerous development partners, also supports Indonesia's REDD+ programs.[211] UN-REDD alone contributed over US$5.6 million to Indonesia between 2009 and 2011.[212]

Moreover, creating strong support for carbon reduction strategies, like REDD+, will lead to increased interest in conservation and sustainable management efforts. Depending upon a number of conditions, oil palm has high value potential relative to other uses of converted land.[213] But studies have shown that conversion of peatlands and forests to agricultural lands 'has generated little economic benefit while releasing substantial amounts of greenhouse gases': in all, Indonesia derives 0.23 Euros per ton of CO_2 from all types of land use on converted land.[214] By comparison, even at current low prices, carbon credits in European markets have traded at about 9.50 Euros per ton of CO_2, and they are expected to rise substantially by 2020.[215] Other studies have estimated that REDD+ payments could offset the costs of expansion into primary forests

---

[208]   World Bank Group, supra note 8, at 18.

[209]   See Asian Development Bank, 'Indonesia Country Strategy and Program 2006–2009 Final Review Validation' (June 2011).

[210]   World Bank Group, supra note 8, at 18.

[211]   Climate Investment Funds, UN-REDD, Forest Carbon Partnership Facility (FCPF), Forest Investment Program (FIP), 'Implementation Progress in Indonesia' Meeting, Washington D.C., 6 Nov. 2010, http://www.climateinvestment-funds.org/cif/sites/climateinvestmentfunds.org/files/Indonesia_Presentation%20 Joint%20Meeting%206%20Nov%202010.pdf (accessed April 16, 2013).

[212]   World Growth, supra note 1, at 19.

[213]   See WBG Synthesis, supra note 29, at 22; Business Watch Indonesia, 'Biofuel Industry in Indonesia: Some Critical Issues' 7 (2007).

[214]   See Rhett Butler, 'Carbon Offset Returns Beat Forest Conversion for Agriculture in Indonesia', Mongabay, 21 Nov. 2007, http://news.mongabay. com/2007/1121-indonesia.html (accessed April 16, 2013) (citing a study by the World Agroforestry Centre, the Center for International Forestry Research, and their Indonesian partners).

[215]   Jeff Coelho, 'More Downside Risk Seen for Hard-hit EU Carbon Market', Reuters, 4 Nov. 2011, http://www.reuters.com/article/2011/11/04/us-carbon-price-analysts-idUSTRE7A35E320111104 (accessed April 12, 2013); Stern Review Report on the Economics of Climate Change, available at http://webarchive. nationalarchives.gov.uk/+/http://www.hm-treasury.gov.uk/independent_reviews/ stern_review_economics_climate_change/stern_review_report.cfm (accessed April 16, 2013) (that estimated present value of damages of one ton of carbon dioxide emitted at US$85).

assuming the price of carbon reaches US$10–33 per ton of CO_2, and US$2–16 per ton for peatland.[216] Even under current carbon prices, the United Nations Environment Programme (UNEP) estimates that avoided deforestation of the peatland forests of Sumatra may be worth up to US$22,000 per hectare in carbon credits over a 25-year period – compared with US$7,400 per hectare if cleared for oil palm plantations.[217] REDD+ facilitates the pricing of carbon by dually harnessing the power of market-based mechanisms and leveraging support from the more developed economies.

REDD+ is not without its own issues, however. For instance, the UN has not yet defined what exactly constitutes a 'forest' protected under REDD+. Indonesia, along with other countries, has contemplated including oil palm plantations and other monoculture developments within that definition.[218] Of course this is unworkable, as rich countries will not invest in programs that effectively fund the destruction of native forests in favor of oil palm conversion. A comprehensive definition protective of natural forests must be articulated within the next year. Relatedly, companies often apply for avoided deforestation credits under REDD+ on forested concessions that they never had any intention of actually converting (e.g. the land was ill-suited for development).[219] Policymakers must guard against this threat – requiring what is referred to as 'additionality' – in order to ensure the most efficient use of limited resources and to prevent inappropriate profiteering.

REDD+ and similar initiatives can put pressure on the palm oil industry as a whole to adopt the management practices necessary to increase sustainability and direct government funds towards capacity building to that end. Transparency and accountability are highly valued by supportive parties in the developed world, but unpredictably achieved. For years, reforestation funds have been hounded by allegations of corruption,

---

[216]   WBG Synthesis, supra note 29, at 22.

[217]   UNEP, 'Orangutans and the Economics of Sustainable Forest Management in Sumatra' (2011) (the report also notes that the carbon value of avoided deforestation in ordinary forests ranges from US$3,711–11,185 per hectare). See also, 'Conservation Offers Better Rewards than Plantations: UNEP', Jakarta Post, 29 Sept. 2011, http://www.thejakartapost.com/news/2011/09/29/conservation-offers-better-rewards-plantations-unep.html (accessed April 12, 2013).

[218]   Jeff Conant, 'Do Trees Grow on Money?', Earth Island Journal, http://www.earthisland.org/journal/index.php/eij/article/do_trees_grow_on_money (accessed April 16, 2013).

[219]   See, e.g., Chris Lang, 'Asia Pulp and Paper's Big REDD Scam on the Kampar Peninsula', REDD-Monitor, 10 Nov. 2011, http://www.redd-monitor.org/2011/11/10/ (accessed April 16, 2013).

mismanagement, and fraud.[220] Therefore, ensuring transparency and measurable progress will be a key component of continued investment in deforestation strategies.

J.   NORWAY-INDONESIA FOREST PROTECTION AGREEMENT

In early 2010, Norway and Indonesia signed a US$1 billion forest protection agreement.[221] The performance-based deal places special emphasis on transparency and progress, tying the distribution of funds to specified benchmarks and emissions targets, spread over three phases of implementation in five years.[222] Moreover, the plan emphasizes alignment with other international initiatives, both bilateral and multilateral, and most crucially REDD+.[223] The deal aims to give all relevant stakeholders – including smallholders and indigenous peoples – the opportunity to effectively participate in REDD+ planning and implementation.[224] Norway further demanded a two-year suspension on all new concessions for the

---

[220]   See, e.g., Rhett Butler, 'Indonesia's Corruption Legacy Clouds a Forest Protection Plan', Yale E360, 27 Dec. 2010, http://e360.yale.edu/feature/indonesias_corruption_legacy_clouds_a_forest_protection_plan/2353/ (accessed April 16, 2013); Transparency International 2007, 'Corruption Perception Index' (2007), www.transparency.org/policy_research/surveys_indices/cpi/2007 (accessed April 16, 2013).

[221]   'Norway to Continue Palm Oil Investments', Mongabay, 30 Mar. 2011, http://news.mongabay.com/2011/0330-norway_fund_palm_oil.html (accesssed April 16, 2013). Norway has made similar commitments for forest protection to Brazil, Ecuador, Guyana, Tanzania, and other countries of the Congo Basin.

[222]   Ewa Krukowska, 'EU CO2 Price Should Be More than Three Times Higher, BNEF Says', Bloomberg, 13 Sept. 2011, http://www.bloomberg.com/news/2011-09-13/eu-co2-price-should-be-more-than-three-times-higher-bnef-says.html (accessed April 16, 2013).

[223]   Kemen Austin, Fred Stolle, and Beth Gingold, 'What's Next for Indonesia-Norway Cooperation on Forests', WRI, 13 July 2010, http://www.wri.org/stories/2010/07/whats-next-indonesia-norway-cooperation-forests (accessed April 16, 2013).

[224]   See Govt. of Indonesia and Govt. of Norway, Letter of Intent between the Government of the Kingdom of Norway and the Government of the Republic of Indonesia on 'Cooperating on reducing greenhouse gas emissions from deforestation and forest degradation,' 2010, http://www.redd-monitor.org/wordpress/wp-content/uploads/2010/05/Norway-Indonesia-LoI.pdf (accessed April 16, 2013) (hereinafter Norway Letter of Intent).

conversion of both peat swamps and virgin rainforests,[225] a moratorium enforced as of January 2011.[226]

Under the first phase of the deal, the two countries will focus preliminarily upon the creation of a REDD+ management agency that will coordinate policies and develop strategies.[227] This national REDD+ office will report directly to the President, and will be responsible for recording progress and identifying both successes and failures.[228] In addition, an independent deforestation monitoring, reporting, and verification (MRV) body will be created.[229] As of yet, neither independent body has been created, resulting in the distribution of only US$30 million by Norway.[230]

Critically, one of the highest priorities of the agreement is the development of an accurate, publicly available database of degraded lands throughout Indonesia.[231] Through such documentation of degraded lands – which are estimated at 23 million hectares of former forest[232] – the government and other public organizations will be able to determine who actually owns and uses these lands that are so critical to the future sustainable development and expansion of the industry.[233] The next two phases will work to improve upon advancements made during the first phase, strengthening mechanisms and enforcing standards – while working towards the development of a long-term climate framework that will substantially reduce both deforestation and emissions in the country.

Early political issues threatened to blunt the effect of the deal.[234] Many

---

[225] Ibid.

[226] Rhett Butler, 'Indonesia's Plan to Save its Rainforests', Mongabay, 14 June 2010, http://news.mongabay.com/2010/0614-indonesia_purnomo_saloh.html (accessed April 12, 2013); Indonesia Climate Change Legislation 3, GLOBE International, http://www.globeinternational.info/wp-content/uploads/2011/04/FINALIndonesia.pdf (accessed April 12, 2013).

[227] World Bank Group, supra note 8, at 18.

[228] Butler, supra note 225.

[229] Butler, supra note 219 (this initiative will likely rely upon satellite and remote sensing technologies that have been used to track deforestation in the Brazilian Amazon).

[230] Amantha Perera, 'Jakarta Must Set Up Forest Bodies to Unblock Aid – Norway', Reuters, 1 Aug. 2011, http://www.trust.org/alertnet/news/jakarta-must-set-up-forest-bodies-to-unblock-aid-norway (accessed April 16, 2013).

[231] See Norway Letter of Intent, supra note 223.

[232] Colchester et al., supra note 11, at 178.

[233] See, e.g., Beth Gingold, 'Degraded Land, Sustainable Palm Oil, and Indonesia's Future', WRI, 13 July 2010, http://www.wri.org/stories/2010/07/degraded-land-sustainable-palm-oil-and-indonesias-future (accessed April 12, 2013).

[234] See, e.g., Butler, supra note 225. After the agreement was signed, provincial

of the specific implementation details of this arrangement have yet to be clarified, and the qualified moratorium established by the deal – which officially covers 64 million hectares but features a long list of exemptions – is not only vague but also inadequate in its duration.[235] Moreover, because the moratorium is a non-legislative document, 'there are no legal consequences if its instructions are not implemented.'[236] Others criticize the failure of the agreement to recognize the rights of indigenous peoples or address existing concessions on forested lands.[237] The government has indicated a willingness to use funds from the deal to compensate developers unable to further convert forests,[238] and to exchange land in return for the cancellation of some forest concessions.[239] However, it has made little progress in the development of inventories and maps, as recent versions fail to identify customary usage, quality of the forest resources, or whether the areas are degraded at all.[240]

Equally disturbing was the signal sent by the government of Indonesia in its decision to turn over more than half of a 91,000-hectare REDD+

---

leaders and governors offered conflicting interpretations of its potential effects. Further, Wandojo Siswanto, head of the climate unit-working group with Indonesia's Forestry Ministry, stated that Indonesia was interested in 'renegotiating' the deal, after some concluded that Norway only agreed to financially support REDD programs, rather than REDD+ (which includes reforestation and sustainable forest management). Later, in October, Siswanto was arrested and charged with taking bribes in exchange for arranging favorable land contracts for a telecommunications company. See Butler, supra note 219.

[235] See, e.g., Deborah Lawrence, 'Indonesia's Forest Moratorium – Analyzing the Numbers', CIFOR, 20 June 2011, http://blog.cifor.org/3272/indonesia%E2%80%99s-forest-moratorium%E2%80%94analyzing-the-numbers/ (accessed April 16, 2013); Alister Doyle, 'Oslo Backs Jakarta's Forest Plan, despite Hurdles', Reuters, 23 June 2011, http://www.reuters.com/article/2011/06/23/us-climate-indonesia-idUSTRE75M43420110623 (accessed April 16, 2013).

[236] CIFOR, 'Indonesia's Forest Moratorium: A Stepping Stone to Better Forest Governance?' (2011). See also Chris Lang, 'New CIFOR Report Points out the Flaws in Indonesia's Forest Moratorium', REDD-Monitor, 1 Nov. 2011, http://www.redd-monitor.org/2011/11/01/ (accessed April 16, 2013).

[237] See, e.g., Chris Lang, 'Norway-Indonesia Forest Deal', REDD-Monitor, 28 May 2010, http://www.redd-monitor.org/2010/05/28/ (accessed April 16, 2013); Adianto Simamora, 'Indigenous Groups Call for Halt to REDD Piilot Project', Jakarta Post, 25 June 2011.

[238] 'Indonesia to Revoke Palm Oil Concession Licenses under Forest Deal', Mongabay, 31 May 2010, http://news.mongabay.com/2010/0531-indonesia_forest_deal.html (accessed April 12, 2013).

[239] See Niluski Koswanage, 'Indonesian Palm Expansion to Halve with Climate Deal', Reuters, 31 May 2010, http://www.reuters.com/article/2010/05/31/us-indonesia-palmoil-idUSTRE64U30720100531 (accessed April 16, 2013).

[240] Perera, supra note 229.

conservation project in Central Kalimantan to palm oil concessions.[241] This debacle reflects deep divisions between the Presidential administration and the Ministry of Forestry, and has already led to a diversion of foreign investments away from the REDD+ programs.[242] Perhaps in reaction to the negative fallout over these concessions and the failure to make progress with Norway, the President expanded the staffing at the country's REDD+ agency.[243] It remains to be seen whether the agreement's potential will ever be fully realized.[244]

K. ASSESSMENT OF SUCCESS OR FAILURE AND PRINCIPAL FACTORS

The overall picture of the palm oil biodiesel program in Indonesia is complicated, though the end result has certainly been less than positive, and undoubtedly more costly than expected. Despite the billions already invested, some have concluded that Indonesia will most likely fail to meet a biofuel standard to which it committed a 5 percent reduction of greenhouse gases by 2025.[245] The biodiesel program was but one iteration in a decade's long push to turn Indonesia into the top palm oil producer in the world. This drive left a vast swath of ecological, social, and economic damages in its wake, coming at great cost to the government, Indonesia's people, and its wildlife. Today, demand for biofuels is growing rapidly but investment in the industry is lagging behind high feedstock and commodity prices and the global credit crisis. This has necessitated ever-higher government subsidization at a time when both aid and revenues are drying up. The Indonesian government has thus poured billions of dollars into

---

[241] Fogarty, supra note 172.

[242] Norman, supra note 64.

[243] 'Indonesia Boosts Staffing of Forest Protection Agency', Reuters, 13 Sept. 2011, http://www.reuters.com/article/2011/09/13/us-indonesia-cliimate-idUS-TRE78C5PW20110913 (accessed April 16, 2013).

[244] See 'Tentatively Favourable Review of Indonesian Delivery on Forest Emission Partnership with Norway', 19 May 2011, http://www.norway.or.id/Norway_in_Indonesia/Environment/Tentatively-favourable-review-of-Indonesian-delivery-on-forest-emission-partnership-with-Norway-/ (accessed April 16, 2013) (Norway's review 'recommends extending the preparatory [first] phase, and use the extra time to build broad support for the forest governance reform agenda').

[245] Frank Swain, 'Indonesia Unlikely to Meet 2025 Biofuel Goal', CIFOR, 8 Sept. 2011, http://blog.cifor.org/4114/indonesia-unlikely-to-meet-2025-biofuel-goal-says-report/ (accessed April 16, 2013).

an enterprise with dubious emissions mitigation benefits and overriding, unqualified environmental harms. The destruction of carbon-rich peatlands and pristine rainforests is irreversible, and yet is only expected to grow over the course of the next half-century – the only question is one of degree.

Many factors combined to result in such discouraging outcomes. The Indonesian government must take primary responsibility, as the predominant central authority for forest and resource management. Persistent corruption in the Ministry of Forestry and other government agencies has interfered with transparency and accountability, and the President, despite expressed intentions, has yet to demonstrate the political leadership necessary to substantially correct these defects.[246] Further, the government has also been complicit in the abuse of the system by corporate interests, electing all too frequently to sanction the illegal backroom deals preferred by the industry. Decentralization of land licensing authority to provincial governments – in the absence of clear federal laws and directives and additional enforcement capacity – was highly detrimental. Extremely poor land use and spatial planning, coupled with dysfunctional cross-sectoral and governmental coordination have further impeded effectiveness. The almost complete lack of environmental considerations and associated safeguards came with predictable results in terms of painful rainforest and peat bog destruction.

Perhaps the government would not have presided over such glaring failures of governance had it not been devoting such enormous resources to fuel subsidization in the first place. It is extremely difficult to manage a program of this magnitude correctly – to guarantee sustainability standards; to enforce environmental compliance over vast distances; to garner broad support from local governments. Incentives and subsidies for both conventional fossil fuels and alternative biofuels are grossly disproportionate when compared to other public spending. This has arguably prevented the government from undertaking the institutional reforms necessary to properly and efficiently manage the transition to more renewable forms of energy. Industry capture is an epidemic, as oil palm, fossil fuel, and timber interests continue to exert undue influence over policymakers. A

---

[246]   For instance, the current forest moratorium fails to specifically include the Ministry of Agriculture or the Ministry of Energy and Mineral Resources within its purview. A recent CIFOR report finds that the 'limited application of the moratorium to activity in these sectors could potentially weaken the government's ability to fulfill the intention of the [moratorium] itself, as well as the President's own commitment to reducing greenhouse gas emissions.' CIFOR, 'Indonesia's Forest Moratorium: A Stepping Stone to Better Forest Governance?' 2 2011.

top-down, comprehensive reform of this system is the only way to achieve the integration of sustainability and thus generate truly sustainable biodiesel production capacity.

L.   LESSONS LEARNED

The Indonesian palm oil program is a classic example of misguided renewable energy development. Virtually everything that could have misfired did. There was a blatant and wholly unacceptable lack of planning from the outset, with no inventorying of appropriate lands, no assessment of environmental risks and no inclusion of the public in the formation of the program. In the rush to profit from converting palm oil into biofuels; to maximize developer profits at the expense of the country, economy, and its environment; in seeking to make Indonesia the world's top biodiesel manufacturer – the project proved to be an unmitigated disaster for all, save the developers.

Most egregiously, the project was an environmental catastrophe, destroying most of Indonesia's forest and peatland and eliminating its rich biodiversity – including the extermination of half the world's rare orangutan population; the decimation of tiger populations by 70 percent to 192 individuals; a decline in elephants of 84 percent, to 210 individuals; and countless others, both known and those still yet to be discovered.

Forest and peatland destruction resulted in a huge increase in greenhouse gas emissions, thus placing Indonesia as the third largest emitter in the world. To reiterate just a few of the most blatant consequences of the project, over 18 million hectares of virgin forest were cleared, ironically most of it for profit from the sale of wood rather than for the creation of palm oil. Illegal logging was found to take place in 37 of 41 Indonesian national parks. The draining of peatlands accounted for 660 million tons of carbon emissions per year, with resulting fires adding another 1.5 billion tons annually. The subsequent greenhouse gas emissions were found to be 800 percent higher than from utilizing conventional biodiesel, and if current expansion plans are effectuated, that figure would increase to 2,000 percent.

Economically, the project proved to be just as colossal of a failure, enriching private entities at the direct expense of the country and its people. The population and its workers suffered disproportionately, and have been saddled with an irreversible legacy of degraded health, natural resources, and economic prospects. Water, air, and land were polluted and damaged. Small landholders were often deprived of their properties. Workers were underpaid and exploited, and no health services were provided. Living conditions were squalid and miserable. Corruption was rampant throughout the implementation of the project.

There is some hope for the future, however, primarily at the instance of international palm oil purchasers. A Roundtable on Sustainable Palm Oil (RSPO) was created to provide standards, criteria, and certification for sustainable palm oil. The European Union (particularly the Netherlands, which had been the largest purchaser) and several other countries, along with multinational corporations like Unilever and Wal-Mart are purchasing only RSPO-certified palm oil. However, in practice too many countries are ignoring the standards. More importantly, there is insufficient funding for oversight and no provision for penalties. Some companies, like Cargill, that proclaimed to be complying evaded the standards through various ruses.

The current Indonesian government is also working with UN agencies to adopt a regime to comply with the REDD+ forest preservation program, providing forest residents with carbon credit funding for forest and peat conservation initiatives. The President of Indonesia announced in 2009 his country's intention to voluntarily cut GHG emissions by 26 percent by 2020 or, with support from developed countries, by up to 41 percent. The government is drafting a REDD+ National Strategy to achieve these goals. The UN REDD+ program contributed over US$5.6 million between 2009 and 2011 to assist with this effort. Unfortunately, there has been fraud perpetrated in this program as well, with companies applying for avoided deforestation credits under REDD+ that they never intended to cut.

Lastly, in 2010 Norway and Indonesia signed a US$1 billion forest protection agreement tying distribution of funds to specified benchmarks and emission targets. The arrangement is designed to give small landholders and indigenous peoples the opportunity to participate in REDD+ planning and implementation. Norway demanded a two-year moratorium on all new concessions of peat swamps and virgin rainforest effective January 2011. The agreement provides for the establishment of a management oversight agency and an independent monitoring, reporting, and verification body. However, neither body has yet been created, and a database of degraded land and land ownership claims has not yet been established. The failure to develop any of these fundamental mechanisms has resulted in a distribution of only US$30 million by Norway thus far. Equally disturbing is a recent agreement by Indonesia to turn over more than half of 91,000 hectares of land in a REDD+ project to palm oil concessionaires.

The bottom line is that although the present Indonesian government has formally recognized the grave social, economic, and environmental consequences of its longstanding palm oil-based biodiesel program, and has made some nominal commitments aimed at rectifying the problems, the prospects for reversing these damages seem to be slight based on performance and subsequent actions. Instead of additional promises, a top-down, comprehensive reform platform coupled with enforcement and meaningful penalties will be required to achieve meaningful progress.

8.   Case study of renewable energy in Pakistan

Richard L. Ottinger with Shakeel Kazmi

A.   INTRODUCTION

Like many other developing countries[1] energy[2] demand in Pakistan is constantly increasing every year. According to the predictions made by the International Energy Agency developing countries will increase their share of global electricity consumption from 20.5 percent in 1999 to 35.8 percent in 2020.[3] The energy need is expected to boost drastically in the future.[4] Approximately 66 percent of the country's energy is generated from fossil

---

[1]   There is no fixed definition of developing countries. The World Bank defines a developing country as: 'A developing country is one in which the majority lives on far less money – with far fewer basic public services – than the population in highly industrialized countries. Five billion of the world's 6 billion people live in developing countries where incomes are usually under $2 per day and a significant portion of the population lives in extreme poverty (under $1.25 per day).' The World Bank, 'What is a Developing Country?', http://web.worldbank.org/WBSITE/EXTERNAL/EXTSITETOOLS/0,,contentMDK:20147486~menuPK:344190~pagePK:98400~piPK:98424~theSitePK:95474,00.html (last accessed 5 Jan. 2013).

[2]   Energy is generally defined as the ability or the capacity to do work and exists in several forms. Energy services are crucial to human well-being and for economic development. In this study the word 'energy' will denote the power generated from fossil fuel or renewable energy sources, used for lighting, heating, cooking, mechanical power, transport, and used for other daily life services.

[3]   Cited in Noel Alter, Shabib Haider Syed, 'An Empirical Analysis of Electricity Demand in Pakistan,' International Journal of Energy Economics and Policy, Vol. 1, No. 4, 2011, pp.116–139, www.econjournals.com/index.php/ijeep/article/download/53/62 (last accessed 5 Jan. 2013).

[4]   'The world will need greatly increased energy supply in the next 20 years, especially cleanly-generated electricity. Electricity demand is increasing twice as fast as overall energy use and is likely to rise 67% from 2010 to 2035.' World Nuclear Association, http://www.world-nuclear.org/info/inf16.html (last accessed 5 Jan. 2013).

fuel.[5] Pakistan is historically a net energy importer; it spent US$ 4.1 billion in the year 2004–05 on crude oil imports.[6] Increasing prices of fuel in the international market[7] and growing concerns of climate change[8] encouraged Pakistan to explore the renewable energy sources.[9] Geographically Pakistan is in a position to exploit wind and solar energy. The country has abundant wind and sun to generate energy.[10] High-speed winds near its major centers have the potential to produce up to 346 gigawatts (GW) of electricity through wind energy.[11] The capacity of solar photovoltaic power in Pakistan is estimated to be 1600 GW per year, enough to provide 3.5kWh of electricity.[12] In March 2007, the president of Pakistan announced that renewable energy should be part of the push to increase energy supplies by 10 to 12 percent every year.[13] The government has also set a target of 10 percent of energy to come from renewable resources by 2015.[14]

The instant study will look at the potential and the barriers that exist to further development of renewable energy. It will address the following

---

[5]   See International Energy Agency, Pakistan: Statistics 2007, http://www.iea.org/stats/countryresults.asp?COUNTRY_CODE=PK&Submit=Submit (accessed 5 Jan. 2013). Approximately 66% of power generation in Pakistan is derived from fossil fuels, followed by hydroelectricity (30%) and nuclear energy (3%).

[6]   See Muhammad Shahid Khalil, Nasim A. Khan, and Irfan Afzal Mirza, 'Renewable Energy in Pakistan: Status and Trends,' Government of Pakistan Ministry of Water and Power, http://www.mowp.gov.pk/gop/index.php?q=a-HR0cDovLzE5Mi4xNjguNzAuMTM2L21vd3AvXNlcmZpbGVzMS9maWxlL3VwbG9hZHMvcHVibGljYXRpY XRppb25zL3JlcGsucGF2L3JlcGsucGF2... 3VwbG9hZHMvcHVibGljYXRpb25zL3JlcGsucGF2... 3VwbG9hZHMvcHVibGljYXRpb25zL3JlcGsucGF2... 3VwbG9hZHMvcHVibGljYXRpb25zL3JlcG... 3VwbG9hZHMvcHVibGljYXRpb25zL3JlcG... 3VwbG9hZHMvcHVibGljYXRpb25zL3JlcG... 3VwbG9hZHMvcHVibGljYXRpb25zL3JlcG... 3VwbG9hZHMvcHVibGljYXRpb25zL3JlcG (last accessed 4 Jan. 2013). Pakistan spent US$ 4.1 billion on crude oil imports for the year 2004–05, which amounts to 25% of the total annual imports for the year.

[7]   '2010 and 2011 have seen a further steady increase in the oil price.' Petrol, www.speedlimit.org.uk/petrolprices.

[8]   See 'Pakistan's total carbon emission make top 25 lists,' United Nations Statistics Division, Department of Economic and Social Affairs, http://mdgs.un.org/unsd/mdg/SeriesDetail.aspx?srid=749.

[9]   Renewable energy denotes energy that comes from natural resources such as sunlight, wind, rain, tides, waves, and geothermal heat.

[10]   See Alternative Energy Development Board (AEDB), 'Wind Energy in Pakistan, Resources Potential,' http://www.aedb.org/wind.htm.

[11]   Ibid. The wind map developed by National Renewable Energy Laboratory (NREL), USA, in collaboration with USAID, has indicated a potential of 346,000 MW in Pakistan.

[12]   See Ibid.

[13]   Triple Bottom-Line, Sustainability Advocacy, 'The Feasibility of Renewable Energy in Pakistan,' http://www.tbl.com.pk/the-feasibility-of-renewable-energy-in-pakistan/. (last visited 4 Jan. 2013)

[14]   A statement by President Musharraf stating that renewable energy should be part of the push to increase energy supplies, ibid.

questions: Why has there been such little development of renewable energy generation? Is slow uptake of renewable energy policies and technologies to blame or a lack of infrastructure and poor competition with conventional power generation? The study intends to identify the strengths and challenges of the future of renewable energy in Pakistan.

B.   PAKISTAN'S BACKGROUND

Geographically the Islamic Republic of Pakistan[15] is a very diverse state. It is divided into three major areas: the Northern highlands; the Indus River plain; and the Baluchistan Plateau.[16] The country covers an area of 803,940 sq. km.[17] It is divided into four provinces: Khyber Pakhtoon Khwa (KPK) (North West Frontier Province (NWFP)),[18] Punjab, Sindh and Baluchistan. The Federal Government manages the Federally Administered Tribal Area (FATA) adjacent to Khyber Pakhtoon Khwa.[19] Azad Kashmir[20] and Northern Areas[21] have their own respective political and administrative machinery.[22] According to the 2008 census the total population of the country is 163,417,500.[23] The reported annual population growth is 1.81 percent.[24] Islam is the official religion of the state. A vast majority of the population (95%) practice Islam.[25] Seventy percent of the population lives in rural areas. Punjab is the most populous province and Baluchistan is the largest in area.

The Central Intelligence Agency (CIA) reports, 'Pakistan remains stuck

---

[15]   The country's official name is Islamic Republic of Pakistan. See the Embassy of Islamic Republic of Pakistan, Facts and Figures, http://embassyofpakistanusa. org/facts&figure1.php.

[16]   See Ibid.

[17]   Embassy of Islamic Republic of Pakistan, Facts and Figures, http:// embassyofpakistanusa.org/facts&figure1.

[18]   Khyber Pakhtoon Khwa (KPK) is the new name of the province. Pakistan's parliament approved the name change from North West Frontier Province (NWFP) to KPK.

[19]   Government of Pakistan, Country Profiles, http://www.infopak.gov.pk/ BasicFacts.aspx. (last accessed 5 Jan. 2013).

[20]   Ibid.

[21]   Pakistan Mission to the UN, http://www.pakun.org/pakistan/. (last visited 5 Jan. 2013).

[22]   Ibid.

[23]   See supra note 17.

[24]   US Department of State, Background Note: Pakistan, http://www.state. gov/r/pa/ei/bgn/3453.htm. (last visited 5 Jan. 2013).

[25]   See supra note 19.

in a low-income, low-growth trap, with growth averaging 2.9% per year from 2008 to 2011'.[26] The UN Human Development Report estimated poverty in 2011 at almost 50 percent of the population.[27] Bloomberg data listed Pakistan as the ninth-poorest country in the Asia-Pacific region with a 2009 gross domestic product per capita of $2609.[28]

Pakistan's climate is generally arid. Summers are hot and winters are cold. A wide variation is reported between extremes of temperature at given locations. There is little rainfall.[29]

C. ENERGY OVERVIEW

Almost 45 percent of Pakistan's population lacks access to electricity.[30] Pakistan currently produces fewer than 14,000 MW, 5000 MW less than its needed requirement. In 2010, 93.35 billion kWh of electricity was produced.[31]

Energy consumption by sector is: Industrial 43 percent, Transport 29 percent, Domestic 20 percent, Commercial 4 percent, Agricultural 2 percent, and Other Government sector 2 percent.[32] The country's per capita energy consumption is 3894 kWh per year compared to the world average of 17,620 kWh. Demand for energy is constantly increasing with the country's rapidly increasing per capita consumption due to growing

---

[26] CIA, The World Fact Book, Pakistan, Energy Production, https://www. cia.gov/library/publications/the-world-factbook/geos/pk.html. (last visited Jan. 6. 2013).

[27] Cited in CIA report, ibid.

[28] Haris Anwer, 'Pakistan Offers Renewable-Energy Incentives to Tackle Shortages,' Aug. 21, 2011, Bloomberg, http://www.bloomberg.com/news/2011-08-25/pakistan-offers-renewable-energy-incentives-to-tackle-shortages.html. (last accessed 3. Jan. 2013).

[29] US Congress, Country Studies, Pakistan, Climate, http://countrystudies.us/pakistan/25.htm. For more details see Pakistan Meteorological Department, Government of Pakistan, http://www.pmd.gov.pk/. (last accessed 4. Jan. 2013).

[30] Dr. Arshad Javed et al., 'Electrical Energy Crisis in Pakistan and Their Possible Solutions,' International Journal of Basic & Applied Sciences IJBAS-IJENS, Vol: 11, No: 05, p. 41, http://www.ijens.org/Vol_11_I_05/110505-9393-IJBAS-IJENS.pdf.

[31] CIA, The World Fact Book, Pakistan, Energy Production, https://www.cia.gov/library/publications/the-world-factbook/geos/pk.html. (last accessed 3. Jan. 2013).

[32] See Dr. Zafar Iqbal Zaidi, 'Renewable Energy Report,' Asian and Pacific Centre for Transfer of Technology (APCTT) recap.apctt.org/download.php?p=Admin/Country/Report/10.pdf. (last accessed 3 Jan. 2013).

population, urbanization trends, development, industrialization, and electrification of rural areas.[33]

Energy is central to the development of the country. Both industrial and agricultural sectors are largely dependent upon the availability of energy. A persistent power crisis has slowed Pakistan's economic activities. Pakistan's president, realizing the link between the development and power, said in a meeting that the gap between the country's power supply and demand was an obstacle to economic and social growth.[34]

A country with a population of over 163 million and a rapidly growing economy has huge energy requirements. Energy needs are becoming acute. A power deficit of 3 to 4 GW a day is triggering 12-hour blackouts that cause riots and close factories in cities nationwide.[35]

Most developing countries, and particularly in the same region as Pakistan like India, Nepal, and Bangladesh, are facing energy shortages, but crises of energy shortage in Pakistan are more crucial. The Economist illustrates the seriousness of Pakistan's energy crises in these words: 'Although Pakistan makes international news for terrorist attacks, anti-American demonstrations and its alleged support for insurgents in Afghanistan, it is the basic inability to switch on a light that is pushing this volatile country closer to the edge.'[36] The Economist further depicts the impacts of electricity outages on people's everyday lives: 'For ordinary people, the frustrations are endless. Refrigerators become useless. Water runs out because it relies on electrical pumps. Children do their homework by candlelight.'[37]

---

[33] 'Electricity consumption in Pakistan has been growing at a higher pace compared to economic growth due to the increasing urbanization, industrialization and rural electrification. From 1970 to the early 1990s, the supply of electricity was unable to keep pace with demand that was growing consistently at 9–10% per annum.' Consulate General of Switzerland in Karachi, Pakistan Power sector, OSEC, Business Network Switzerland, p.2, http://www.osec.ch/de/filefield-private/files/26090/field_blog_public_files/5513. (last accessed 4 Jan. 2013).

[34] Syed Fareed Hussain, 'Powering up: Pakistan's Push for Renewable Energy,' SciDev Net, Jun. 06, 2007, http://www.scidev.net/en/features/powering-up-pakistans-push-for-renewable-energy.html. (last accessed Jan. 3. 2013).

[35] Haris Anwar, 'Pakistan Offers Renewable-Energy Incentives to Tackle Shortages,' Bloomberg, 25 Aug 2011, http://www.bloomberg.com/news/2011-08-25/pakistan-offers-renewable-energy-incentives-to-tackle-shortages.html. (last accessed 5 Jan. 2013).

[36] Print News, 'Pakistan's Energy Shortage, Lights Out, Another Threat to a Fragile Country's Stability,' The Economist, Oct 8th 2011, Islamabad, http://www.economist.com/node/21531495. (last visited 05 Jan. 2013).

[37] Ibid.

D. ENERGY MIX

Pakistan's optimum capacity to generate electricity in summer is 17,897 MW, which is reduced to 13,215 MW in winter.[38] Public and private sectors are involved in energy generation.[39] Primary energy sources in Pakistan are oil, gas, coal, hydro, and nuclear electricity.[40] According to the International Energy Agency report in 2007 approximately 66 percent of power generation in Pakistan is derived from fossil fuels (primarily oil and gas) followed by hydroelectricity (30%) and nuclear energy (3%).[41] The Pakistan Meteorological Department reports in its Energy Project that oil accounts for approximately 45 percent, gas 34 percent, hydropower 15 percent, coal 6.2 percent, and nuclear 2.34 percent[42] of the total commercial energy supply.[43] The Ministry of Petroleum has listed the following sources of energy in Pakistan: oil 23.3 percent, gas 51.6 percent, LPG 0.4 percent, coal 6.2 percent, hydroelectricity 11.3 percent, and nuclear electricity 1.2 percent.[44]

Energy production is largely reliant on fossil fuels that are largely imported. Pakistan's internal oil production meets only one sixth of the country's current oil requirements.[45] The power and industrial plants were

---

[38] See Saif Ullah and Arif Allaudin, 'Quantum Leap in Wind Power in Asia, Structure Consultation, Opportunities & Challenges to Scaling Up Wind Power in Pakistan,' presentation, Pakistan Power Sector Overview, 21–22 June 2012, ADB Headquarters, Manila, http://www.sari-energy.org/PageFiles/What_We_Do/ activities/Quantum_Leap_in_Wind_Power_in_Asia-Manila_Conference/presen tations/ALLAUDIN_-_Presentation_Manila.pdf. (last accessed 5 Jan. 2013).

[39] Fifty-nine percent of energy is produced by the public sector and 41% by the private sector, ibid.

[40] See Ullah, supra note 38.

[41] International Energy Agency, http://www.iea.org/index_info.asp?id=2401. (last visited 4 Jan.5 2013).

[42] The Pakistan Atomic Energy Commission is successfully operating three nuclear power plants (ANUPP near Karachi, Chasma-1 and Chasma-2 in Punjab), see Pakistan Atomic Energy Commission (PAEC), http://www.paec.gov.pk/ paec-np.htm (last visited 23 Mar. 2012). Pakistan has a small nuclear power program, with 725 MWe capacity, see World Nuclear Association, 'Nuclear Energy in Pakistan,' http://www.world-nuclear.org/info/inf108.html.

[43] Pakistan Meteorological Department, Energy Project, http://www.pakmet. com.pk/wind/Wind_Project_files/Page767.html. (last accessed 3 Jan. 2013).

[44] Cited in Muhammad Shahid Khalil, Nasim A. Khan, and Irfan Afzal Mirza, 'Renewable Energy in Pakistan: Status and Trends,' http://202.83.164.28/ mowp/userfiles1/file/uploads/publications/repk.pdf. (last accessed 4 Jan. 2013).

[45] See Dr. Zafar Iqbal Zaidi, 'Renewable Energy Report,' Asian and Pacific Centre for Transfer of Technology (APCTT). Natural gas production during 2007–08 was 3,973 million cubic feet per day and oil production was 69,954

converted to gas to reduce oil imports. The bulk consumption of gas in industry, increasing domestic use, and use as fuel in motor vehicles has created a serious shortage of gas in the country. Power shortage related crises are intensifying because of a decline in the drilling activities and slow pace of the government's plans to import gas (from Iran through a pipeline and liquefied natural gas (LNG) from Qatar).

E. GOVERNMENT ENERGY STRUCTURE

Various departments and ministries are responsible for energy management. Principal players include: the Water and Power Ministry,[46] the Ministry of Petroleum and Natural Resources,[47] the National Electric Power Regulatory Authority (NEPRA),[48] the Private Power and Infrastructure Board (PPIB),[49] the Karachi Electric Supply Corporation (KESC),[50] the Water and Power Development Authority (WAPDA),[51] and the National Transmission and Dispatch Company (NTDC).[52]

---

barrels per day. During 2007–08, 27 exploratory and 53 development wells were drilled, mostly of gas/condensate, out of which 5 were by Oil & Gas Development Corporation Limited (OGDCL) and 6 by other companies.

[46]   See Ministry of Water and Power, Government of Pakistan, http://www. mowp.gov.pk/.

[47]   Ministry of Petroleum and Natural Resources, Government of Pakistan, http://www.mpnr.gov.pk/ (last visited 5 Jan. 2013).

[48]   The National Electric Power Regulatory Authority (NEPRA) was set up under the Regulation of Generation, Transmission and Distribution of Electric Power Act 1997 (known as the 'NEPRA Act'). It is mandated to act as an independent regulator for the provision of electric power services in Pakistan.

[49]   A Private Power and Infrastructure Board (PPIB) was established as a 'one window' facilitator for conventional private power sector generation projects, including hydro projects of above 50 MW capacity.

[50]   The Karachi Electric Supply Corporation (KESC) provides distribution services under license from NEPRA.

[51]   Water and Power Development Authority (WAPDA), http://www.wapda. gov.pk/. (last visited 3 Jan. 2013).

[52]   National Transmission and Dispatch Company (NTDC), a sole transmission system operator. NTDC is licensed by NEPRA. NTDC transmits power purchasing through the Pakistan Power Holding Company Ltd., and the Central Power Procurement Agency (CPPA) NTDC is the System Operator for the secure, safe, and reliable operation, control, and dispatch of generation facilities as well as the Transmission Network Operator for the operation and maintenance, planning, design, and expansion of the national transmission network.

F. ROLE OR RENEWABLE ENERGY

Shortage of power is rising due to growing energy demand. The oil import prices are soaring and locally available gas reserves are depleting rapidly.[53] Recently alternate energy has grabbed the attention of the policy makers and researchers in Pakistan to explore the other energy options.[54] Renewable energy resources are widespread and abundant in Pakistan, and the government of Pakistan is looking to exploit the potential in wind, hydro, solar, and biomass.[55] According to an estimate wind energy could produce 350,000 MW, solar 2.9 million MW, and geo-thermal 2,500 MW in Pakistan.[56]

1. Renewable Energy Resources

All the major renewable energy resources are common and bountiful in Pakistan. The government of Pakistan and experts believe that available energy resources have the potential to meet the country's energy needs. This section presents an overview of the potential and available renewable sources in the country.

a. Solar

Pakistan's sunny climate offers a very good opportunity for the solar production of energy.[57] Pakistan lies in a region of high solar irradiance. The country receives about 15.5×1014 kWh of solar irradiance each year.[58] There are only a few cloudy days in most of the region. The

---

[53] Natural Gas Asia, 'Pakistan Depletes Half of Its Gas Reserve,' 27 Dec. 27, 2011, http://www.naturalgasasia.com/pakistan-depletes-half-of-its-gas-reserves-4030. (last visited 4 Jan. 2013).

[54] Between 1983 and 1988 the government of Pakistan invested 14 million rupees in feasibility studies for solar energy and biogas production, but no pivotal project developments resulted.

[55] Statement by the Prime Minister of Pakistan in a meeting of the Alternative Energy Development Board (AEDB) chaired by him, to review the Renewable Energy Development Projects, Staff Report, Daily Times, Mar. 7, 2012, http://www.dailytimes.com.pk/default.asp?page=2012%5C03%5C07%5Cstory_7-3-2012_pg5_6 (last accessed 16 Mar. 2012).

[56] See ibid.

[57] Mashael Yazdanie, 'Renewable Energy in Pakistan: Policy Strengths, Challenges & the Path Forward,' Energy Economics & Policy, Thomas Rutherford (2010). Also see, Dr. Muhammad Shahid Khalil et al, infra note 60.

[58] Alternative Energy Development Board, Ministry of Water and Power, Government of Pakistan, 'Resource Potential of Solar Photovoltaic,' http://www.aedb.org/SolarPV.htm. (last accessed 21 Dec. 2012).

average amount of daily sunlight in Pakistan is nine and a half hours.[59] The capacity of solar photovoltaic power is estimated to be 1600 GW per year, enough to provide 3.5 kWh of electricity.[60] The mean global irradiation falling on horizontal surface is about 200–250 W per m2 per day. This amounts to about 1500–3000 sunshine hours and 1.9–2.3 MWh per m2 per year. The Solar Energy Research Centre (SERC) in its report suggests that solar energy is very much feasible for Pakistan. A report suggests that if only 0.25 percent of Baluchistan's land is covered with solar panels it will generate enough electricity to meet all of the country's energy demand.[61]

Seventy percent of Pakistan's population lives in rural settings far away from the major centers.[62] It can be much cheaper to provide remote local areas with solar energy than connecting them with long distance grids.

Presently there is no major solar energy distributed through a commercial grid operational in the country. Still, solar panels are installed in some rural areas of the country to generate the power to run an electric fan, light bulbs, and to generate the electricity to extract water for a drip irrigation system. Most of the solar technology is imported from other countries. Only one factory located at Hattar Industrial Area near Taxila has started manufacturing solar panels in Pakistan.

b.   Wind energy

Wind energy is a very useful renewable energy source. It is pollution-free and it does not create greenhouse gasses, toxic or radioactive waste. It is a sustainable source of energy so long as there is sufficient wind available.[63]

Wind energy is not a new invention. Its history goes back to AD 500 when it was first used to grind grain and pump water in Persia.[64] In Europe

[59]   Ibid.

[60]   See Dr. Muhammad Shahid Khalil, Dr. Nasim A. Khan, and Irfan Afzal Mirza, 'Renewable Energy in Pakistan: Status and Trends,' http://202.83.164.28/mowp/userfiles1/file/uploads/publications/repk.pdf (last accessed 22 Dec. 2012): 'The mean global irradiation falling on horizontal surface is about 200–250 watt per m2 per day. This amounts to about 1500–3000 sun shine hours and 1.9–2.3 MWh per m2 per year.'

[61]   Triple Bottom-Line, Sustainability Advocacy, 'Solar Energy: A Feasible Alternative for Pakistan,' http://www.tbl.com.pk/the-feasibility-of-renewable-energy-in-pakistan. (last accessed 4 Jan. 2013).

[62]   According to a report by the Solar Energy Research Centre (SERC) 70% of the population lives in 50,000 villages that are very far away from the national grid in Pakistan, ibid.

[63]   Pakistan Meteorological Department, Energy Project, http://www.pakmet.com.pk/wind/Wind_Project_files/Page767.html. (last visited 22 Dec. 2013).

[64]   Darrell M. Dodge, 'Illustrated History of Wind Power Development,' TelosNet, http://telosnet.com/wind/early.html (last accessed 5 Jan. 2013).

the Dutch have used windmills since the eleventh century.[65] Recently, the concept of wind energy has been widely used in Europe and North America, and China and India are aggressively increasing their use of wind energy technology. China today is the world's leader in wind machine manufacture.

Pakistan is one of the countries with high wind speeds. Wind speed is very high near major centers including Karachi, the largest city, and Islamabad, the capital city of Pakistan.[66] The wind speed near Karachi ranges between 6.2 and 6.9 meters per second and near Islamabad, the wind speed is anywhere from 6.2 to 7.4 meters per second. In the Baluchistan and Sindh provinces, ample wind exists to power every coastal village in the country. Gharo and Keti Bandar corridors have enough wind to produce between 40,000 and 50,000 MW of electricity.[67]

The ideal location for wind turbines is an area where an uninterrupted flow of wind is available. Areas in or near water often enjoy an uninterrupted flow of wind. Pakistan's geography provides ample areas to install wind turbines in or near water as Pakistan has several rivers and lakes.

c. Hydro and biomass energy

Hydro is a truly sustainable energy resource in Pakistan's energy mix. At present Pakistan has an installed hydropower capacity of approximately 6.6 GW. Pakistan is using only 16 percent of its available potential of 41.5 GW hydro energy.[68] Hydro is 15 percent of Pakistan's energy generation mix.[69] Three hundred sites have the potential to generate approximately 350 MW of electricity. A memorandum of understanding between the government of Pakistan and the Turbo Institute of Slovenia has been signed for the transfer of technology to make micro hydro turbines.[70]

Biomass converted to biofuels[71] and that derived from waste comprise

---

[65]   Ibid.

[66]   Alternative Energy Development Board, Ministry of Water and Power, Government of Pakistan, 'Wind Energy in Pakistan, Resources Potential,' http://www.aedb.org/wind.htm. (last visited 4 Jan. 2013).

[67]   Ibid.

[68]   Harijan, Uqaili, and Memon (2008), cited in Mashael Yazdanie, 'Renewable Energy in Pakistan: Policy Strengths, Challenges & the Path Forward, Hydro Energy,'   http://www.cepe.ethz.ch/education/termpapers/Yazdanie.pdf.   (last accessed 22 Dec. 2013).

[69]   Ibid.

[70]   See Khalil, supra note 6.

[71]   IUCN, 'What Are Biofuels?', Fact sheet on Biofuels, World Conservation Congress, 2008, http://cmsdata.iucn.org/downloads/biofuels_fact_sheet_wcc_30_sep_web.pdf. (last accessed 4 Jan. 2013). 'Biofuels are liquid fuels derived from

solid biofuels, liquid biofuels, biogases, industrial waste, and municipal waste.[72] Biomass derived fuels have the potential to generate 6.6 percent of Pakistan's current power generation. Seventy percent of Pakistan's population lives in rural areas and approximately 48 percent of the total population farm for their living. Approximately 50,000 tons of solid waste, 225,000 tons of crop residue, and over 1 million tons of animal manure are estimated to be produced daily.[73] The potential production of biogas from livestock residue is 8.8 to 17.2 billion cubic meters of gas per year, which is equivalent to 55 to 106 TWh of energy. The annual electricity production from biofuels is estimated at 5700 GWh.[74]

G. RENEWABLE ENERGY REGIME

In the last 30 years, the government of Pakistan has established several institutions and associations to promote renewable technologies to generate power. The Alternative Energy Development Board (AEDB), and the Pakistan Council of Renewable Energy Technology (PCRET)[75] are considered key players. The SAARC Energy Center Islamabad, the South

---

non-fossil biomass (recently living organisms and their metabolic by-products).' While biofuels are generally thought of as vehicle fuels, they can be used in any application that currently uses liquid fuels, e.g. in generators or cooking stoves. Biogas is also sometimes included as a biofuel when it is used in engines instead of liquid petroleum gas (LPG) or compressed natural gas (CNG). Currently, the two main biofuels are ethanol and biodiesel.

[72]   International Energy Agency, Glossary, http://www.iea.org/textbase/nppdf/ free/2011/ key_world_energy _stats.pdf (last visited 21 Mar. 2012). Included as biofuels are wood, vegetal waste (including wood waste and crops used for energy production), ethanol, animal materials/wastes, and sulphite lyes. Municipal waste comprises wastes produced by residential, commercial, and public services that are collected by local authorities for disposal in a central location for the production of heat and/or power.

[73]   Harijan, Uqaili, and Memon (2008), cited in Mashael Yazdanie, 'Renewable Energy in Pakistan: Policy Strengths, Challenges & the Path Forward,' http://www. cepe.ethz.ch/education/termpapers/Yazdanie.pdf. (last accessed 22 Dec. 2012).

[74]   Ibid.

[75]   The National Institute of Silicon Technology (NIST) was established in 1981. The primary purpose was to do research and promote development in the field of solar energy. The Pakistan Council for Appropriate Technology (PCAT) was organized in 1985 to promote hydropower, biogas, and small-scale wind technologies. In 2002 both institutions were officially merged and renamed as the Pakistan Council of Renewable Energy Technology (PCRET). The stated objective of PCRET remained organization, coordination, and promotion of renewable energy.

Asia Initiative in Energy (SARI/E), and the UN Development Program (UNDP) are major regional and international organizations involved in promoting renewable energy in Pakistan.

1. Alternative Energy Development Board (AEDB)

In May 2003, the government of Pakistan formed the Alternative Energy Development Board (AEDB).[76] Prior to the establishment of AEDB there was no agency for the planning and development of renewable energy. The main object of the AEDB is to promote the cleaner power from renewable energy sources. The AEDB acts as the central national body for alternate and renewable energy. The main objective of the AEDB is to facilitate, promote and encourage the development of renewable energy in Pakistan.[77] The AEDB is additionally charged with providing electricity services to 7,874 remote villages located far from the national power grids in Sindh and Baluchistan provinces of Pakistan.[78] The AEDB has introduced renewable energy policies, has launched various renewable energy

---

[76] Alternative Energy Development Board, Ministry of Water and Power, Government of Pakistan, 'Wind Energy in Pakistan,' http://www.aedb.org/Ordinance.htm. (last accessed 20 Mar. 2012).

[77] AEDB, Terms of Reference, http://www.aedb.org/TermsOfReference.htm (last visited 20 Mar. 2012). The terms of reference are listed as following:

1. To develop national strategy, policies and plans for utilization of alternative and renewable energy resources to achieve the targets approved by the Federal Government in consultation with the Board. 2. To act as a forum for evaluating, monitoring and certification of alternative or renewable energy projects and products. 3. To facilitate power generation through alternative or renewable energy resources by: a) Acting as one window facility for establishing, promoting and facilitating alternative or renewable energy projects based on wind, solar, small-hydro, fuel cells, tidal, ocean, biogas, biomass, etc. b) Setting up alternative and renewable energy power pilot projects on its own or through joint venture or partnership with public or private entities in order to create awareness and motivation of the need to take such initiatives for the benefit of general public as well as by evaluation concepts and technologies from technical and financial perspective. c) Conducting feasibility studies and surveys to identify opportunities for power generation through alternative and renewable energy resources. d) Undertaking technical, financial and economic evaluation of the alternative or renewable energy proposals as well as providing assistance in filing of required licensing applications and tariff petitions to National Electric Power Regulatory Authority (NEPRA). e) Interacting and coordinating with the National and International agencies for promotion and development of alternative energy. f) Assisting the development and implementation of plans with concerned authorities and provincial Governments for off-grid electrification.

[78] Ibid.

projects and has acquired 18,000 acres of land suitable for the installation of wind turbines in the future.[79]

2. International Technical and Financial Support

The South Asian Association for Regional Cooperation (SAARC) Energy Center in Islamabad and the South Asia Initiative in Energy (SARI/E) were established after their endorsement in the SAARC thirteenth annual summit in October 2005. The goals of the Center are to strengthen South Asia's capacity to collectively address the regional and global energy issues, to facilitate energy trade within the SAARC region and to enhance more efficient use of energy within the region.[80]

The Productive Use of Renewable Energy (PURE),[81] a UN Development Program (UNDP) agency, has aided the Alternative Energy Development Board to promote renewable energy. UNDP sponsors projects related to development of wind power, conservation, and technologies related to energy efficient housing and the training of women to use household appliances.

The USAID program has provided technical assistance on Pakistan's LNG import policies and provided small grants for solar lamps and solar-powered pumps in Baluchistan. USAID offered support for establishing the SAARC Energy Center in Islamabad. Moreover, it helped in the preparation of a wind and solar atlas for Pakistan.[82]

Germany has financed some projects to strengthen the transmission system and has been looking into small and medium size hydropower plants and has been compiling and carrying out analytical studies of them. Germany also has been supporting capacity-building at the Alternative Energy Development Board.

The Canadian International Development Agency (CIDA) provided support to the Pakistani Water and Power Development Agency (WAPDA) to rehabilitate Warsak Hydroelectric Power Station. CIDA also provided

---

[79]   Ibid. Visit AEDB to see a complete list of projects, http://www.aedb.org/Ordinance.htm. (last accessed 22 Dec. 2012)

[80]   SAARC Energy Center, Islamabad, http://www.saarcenergy.org/. (last accessed 4 Jan. 2013).

[81]   UNDP Pakistan, http://undp.org.pk/productive-use-of-renewable-energy-pure.html. Also see http://undp.org.pk/images/documents/Pure%20Pro%20Doc.pdf. (last accessed 4 Jan. 2013).

[82]   Mashael Yazdanie, 'Renewable Energy in Pakistan: Policy Strengths, Challenges & the Path Forward,' Energy Economics & Policy, Thomas Rutherford (2010).

assistance to strengthen capacity at the Ministry of Petroleum and Natural Resources and for the development of sound policies and regulatory frameworks to encourage private sector investment, as well as to enforce effective environmental protection and the sound management and conservation of hydrocarbon resources.

The government of Japan is considering financing the modernization of the national power dispatch center in concert with an Asian Development Bank (ADB) transmission loan. Japan has also expressed an interest in Pakistani rural electrification.

A Dutch non-governmental organization, SNV, has looked into supporting the deployment of biogas systems in rural areas to utilize livestock waste. SNV recently completed a study on this subject with Winrock International and the UNDP.

Where international financial support is concerned, the Asian Development Bank (ADB) is currently the major investor in Pakistan's electricity sector.[83] ADB is leading all international financial institutions with a \$3.5 billion pipeline of loans and three major new investment loans. The Renewable Energy Development Sector Investment Program of ADB will loan up to \$510 million to support the development of renewable energy resources in Pakistan. The program will finance run-of-the-river small hydro plants in Punjab and KPK, and investments in wind, solar, and biomass projects.[84]

H. RENEWABLE ENERGY POLICIES

Renewable energy development was virtually non-existent in the past. All energy policies implemented between 1985 and 2002 emphasized the need for employing renewable energy resources but none provided a framework for the implementation of renewable projects.[85] These policies failed to attract the private sector's confidence and investment.[86] In the last decade the government began to intensify its efforts to stimulate the

---

[83] Pakistan: Renewable Energy Development Sector Investment Program, WRI, http://projects.wri.org/sd-pams-database/pakistan/renewable-energy-devel opment-sector-investment-program. (last visited 4 Jan. 2013).

[84] See Asian Development Bank, http://www.adb.org/Media/Articles/2006/ 11118-Pakistan-renewable-energy-d. (last visited 22 Dec. 2012).

[85] Khattak et al. (2006). Cited in Mashael Yazdanie, 'Renewable Energy in Pakistan: Policy Strengths, Challenges & the Path Forward, Energy Policy in Pakistan,' http://www.cepe.ethz.ch/education/termpapers/Yazdanie.pdf. (last accessed 4 Jan. 2013).

[86] Ibid.

renewable industry, with various policies, incentives, and initiatives. The Prime Minister of Pakistan revealed at an AEDB meeting that he chaired that the country possessed an enormous potential of renewable energy power generation.[87] He announced that wind energy has the potential to produce 350,000 MW, solar 2.9 million MW and geothermal 2500 MW of renewable energy in Pakistan.[88]

I.   POLICY FOR DEVELOPMENT OF RENEWABLE ENERGY FOR POWER GENERATION (2006)

In 2006 Pakistan launched its first renewable energy policy to explore its untapped renewable energy resources such as wind, solar, and biomass.[89] The primary aim of the government of Pakistan's 'Policy for Development of Renewable Energy for Power Generation' was to adopt renewable energy technologies to cope with the continuously growing energy short-age.[90] The government of Pakistan also claimed that the purpose of the new renewable energy policy was to promote environmental protection and awareness.[91] The renewable energy policy proposed the relaxation of governmental approvals and requirements for renewable projects.

Other salient features of the renewable energy policy included: to increase the deployment of renewable energy technologies, to promote private sector investment, easy financing, to encourage the domestic manufacturing industry, and to promote environmental protection and awareness. The policy exempted non-IPP projects[92] from approval and allowed all to put up their own projects. There is no income tax on renewable energy projects. All renewable energy equipment is free of sales tax and customs duties.[93] The government's Policy for Development of Renewable Energy for Power Generation offers wind risk insurance in certain areas (risk of variability of wind speed) and provides for guaranteed electricity purchase; grid access and protection against political risk are the responsibility of the purchaser.

---

[87]   'New Energy Policy to Attract Investors,' Staff Report, Daily Times, 15 Oct. 2011, http://www.dailytimes.com.pk/default.asp?page=2011%5C10%5C15% 5Cstory_15-10-2011_pg5_10. (last visited 22 Dec. 2012).

[88]   Ibid.

[89]   AEDB introduced the Policy for Development of Renewable Energy for Power Generation in 2006.

[90]   See Yazdanie, supra note 82.

[91]   Ibid.

[92]   See supra note 87. Independent Power Producer or IPP projects are for selling power to the grid exclusively.

[93]   Ibid.

It also offers an attractive tariff (cost plus with up to 17% Return On Equity (ROE)) indexed to inflation and exchange rate variation (rupee/ dollar). Euro/dollar parity is allowed and possible carbon credits are available. Issuance of corporate registered bonds is also allowed.[94]

J. ALTERNATIVE AND RENEWABLE ENERGY POLICY (ARE 2011)

The Alternative Energy Development Board of Pakistan (AEDB) in consultation with the federal Ministry of Water and Power and other governmental and non-governmental bodies drafted the recently announced Alternative and Renewable Energy Policy (ARE 2011).[95] The new policy sets the target of at least 5 percent of total commercial energy supplies through alternative and renewable energy sources by 2030.[96] The policy states in its introduction: 'ARE Policy 2011 provides a comprehensive framework encompassing wider scope for utilization of all ARE sources; not only for the purposes of generation of electricity but also for encouraging recourse and utilization of ARE technology (ARET)-based applications by commercial and domestic consumers.' Imran Ahmed, the policy director of AEDB, stated that the primary difference between the new and the 2006 policy is that the focus of the previous renewable energy policies was wind and hydroelectric power, but the new policy includes other

[94] Ibid.
[95] In a presentation at the Asia Development Bank's headquarters in Manila, AEDB's chair highlighted the following key aspects of the ARE Policy 2011: 1-Wind risk taken by the government, 2-Guaranteed electricity purchase 3-Grid provision is the responsibility of the purchaser, 4-Attractive cost of land for wind energy projects, 5-Standardized project documents (LOI, LOS, AI, EPA, etc), 6-Attractive tariff (return on investment 17% to 18%), 7-No import duties on equipment, 8-Zero sales tax, 9-No income tax/withholding tax, 10-Repatriation of equity along with dividends freely allowed, 11-Permission to issue corporate registered bonds, 12-Easy convertibility of PKR into USD, 13-Net metering, 14-Banking of electricity, 15-Wheeling provisions, 16-Grid spill over concept introduced, 17-Carbon credits mainly to the investor. See Saif Ullah and Arif Allaudin, 'Quantum Leap in Wind Power in Asia, Structure Consultation, Opportunities & Challenges to Scaling Up Wind Power in Pakistan,' presentation, Pakistan Power Sector Overview, June 21–22, ADB Headquarters, Manila, http://www.sarienergy.org/PageFiles/What_ We_Do/activities/Quantum_Leap_in_Wind_Power_in_Asia-Manila_Conference/ presentations/ALLAUDIN_Presentation_Manila.pdf. (last accessed 4 Jan. 2013).
[96] See Alternative and Renewable Energy Policy (ARE 2011), Alternative Energy Development Board, Ministry of Water and Power, Government of Pakistan, Wind Energy in Pakistan, http://www.aedb.org/Ordinance.htm. (last visited 5 Jan. 2013).

sources of renewable energy such as geothermal, ocean waves and tides, solar, and bio-waste.[97]

Net metering and wheeling are salient features of this policy. The new renewable energy policy allows producers net metering and billing.[98] Section 4.4.1 of the policy provides for wheeling.[99] IPP producers are also permitted to inject electricity at one point on the one grid and receive an equivalent amount at another location.[100]

This policy attempts to improve incentives over the earlier 2006 policy for individual consumers installing solar panels in their homes and for larger investors. ARE 2011 provides the approved tariff to ensure that the electricity is priced competitively, with the rate dropping over time as project debt is paid down. The government is set to raise feed-in tariffs (FITs) requiring electricity supply companies to purchase electricity that guarantees up to 18 percent return to private producers of wind and solar power.[101]

Section 4.1.1 guarantees the market for investors and makes it mandatory for electricity companies to purchase the power from renewable

---

[97]   Statement by Imran Ahmed, Policy Director AEDB, cited in Sleem Shaikh, Sughra Tunio, 'Pakistan to Boost Renewable Energy to Meet Power Shortfall,' Alertnet, 6 Jul. 2011, http://www.trust.org/alertnet/news/pakistan-to-boost-renewable-energy-to-meet-power-shortfall/. (last visited 4 Jan. 2013).

[98]   Net metering and billing allows you to sell surplus electricity to the grid at a given point in time and draw electricity, as required, at a later point in time. Section 4.4.3.1 Net-Metering of the proposed ARE 2011 reads, 'A consumer based ARE DG up to 1 MW has an option to sell full or part of generated electricity to the grid, which is netted against the energy delivered by the grid. Under this mechanism utility consumers are encouraged to generate their own electricity from ARE resources. Under this arrangement the tariff charged would be the applicable retail tariff to the premise (e.g., industrial, commercial, or residential rates)'.

[99]   See section 4.4.1 Wheeling, Alternative and Renewable Energy Policy (ARE 2011), supra note 96. Wheeling means the operation whereby distribution system and associated facilities of a transmission licensee or a distribution licensee are used by another person for the conveyance of electricity on payment of charges to be determined. According to the policy, renewable energy power producers shall also be allowed to enter into direct (bilateral) sales contracts with end-use customers. Under this arrangement, they would be allowed to sell all or a part of the power generated by them to their direct customers, and the rest to the utility for general distribution. For direct sales, they shall be required to pay 'wheeling' charges for the use of the transmission and/or distribution grid network used to transport the power from the plant to the purchaser.

[100]   See Ibid. Producers are charged for accounting known as a wheeling charge.

[101]   See section 4.3.3 Feed in Tariffs, ARE 2011, 'GOP recognizes that Feed-in Tariff has been globally tested tool to attract prompt investment in ARE sector. Feed-in tariffs are therefore to be announced by NEPRA in respect of various ARE sources at such levels as deemed appropriate and duly supported by relevant NEPRA rules on the subject'.

projects. The section reads, 'It shall be mandatory for NTDC/CPPA/ DISCOs [National Transmission Dispatch System/Central Power Purchase Agency/Distribution companies] to buy all the electricity offered to them by ARE projects established pursuant to the ARE Policy 2011 at rates determined by NEPRA.'[102] The policy simplifies the licensing procedure for IPPs up to 5 MW and requires a concessional fee structure.[103]

K. DESCRIPTION OF RENEWABLE ENERGY PROJECTS

The present government has demonstrated a serious interest in promoting electricity from renewable sources, especially solar and wind. The government has initiated a number of projects in the power sector to overcome the energy shortages in the country,[104] but most of the solar or wind power projects for exploitation on a commercial basis are not operational yet.[105] This section will look at some of the non-commercial solar projects and wind power commercial projects.

1. Solar

In the village of Narian Khorian a local firm installed 100 solar panels. The village is only 50 kilometers away from Islamabad.[106] The homeowners had no electricity in their homes prior to installation of these panels. These solar panels generate sufficient power to run an electric fan and light bulbs.[107]

---

[102] Section 4.1.1.1 Guaranteed Market: Mandatory Purchase of Electricity, ARE 2011, supra note 96.

[103] See Section 4.2.1, 'Simplified Generation Licensing Procedure,' ARE 2011, supra note 96.

[104] See statement by Prime Minister Syed Yusuf Raza Gilani, The News, Our Correspondent, Wednesday, March 07, 2012. 'Govt. has started many power projects: PM,' http://www.thenews.com.pk/Todays-News-2-96465-Govt-has-started-many-power-projects-PM. (last visited 5 Jan. 2013).

[105] On 24 December the President of Pakistan, Asif Ali Zardari, inaugurated the first wind project in the country, 'President to Inaugurate Jhimpir Wind Farm Today,' ARY News, 24 December 2012, http://www.arynews.tv/english/newsdetail.asp?nid=68137 (last visited 29 Dec. 2012).

[106] Azfar A. Khan, 'Energy,' Utilizing Solar in Pakistan, Chitral News, http://www.chitralnews.com/Solr%20systems-12--Oct-2011.htm (last visited 5 Jan. 2013). Also see AEDB, http://www.aedb.org. (last visited 29 Dec. 2012).

[107] Ibid. Khan, Utilizing Solar in Pakistan.

The AEDB has completed a solar project. The villagers were given solar panels and solar cookers. The project resulted in environmental and economic benefits such as an 80 percent decrease in deforestation in the project area. And because the solar cookers were manufactured in Pakistan, it generated local economic growth.

In another program by the Thardeep Rural Development Programme (TRDP) solar panels were provided in Nagarparkar (a village in the province of Sind called Wandhan jo Wandio) to generate the electricity to extract water for a drip irrigation system.

In November 2010, the Society for Conservation and Protection of Environment (SCOPE), a non-governmental organization, installed solar panels to provide electricity to 45 households in the village of Gul Hassan Shoro, in the Thatta District. The solar power project lit people's homes and revolutionized their lifestyle, as there was no electricity in the village prior to this project.

2.   Wind Energy

The government is convinced that wind energy is a viable source of energy for the country. It has installed 85 micro turbines in Mirpur Sakro. These turbines are enough to power 356 homes. Forty turbines installed in Kund Mali have the capacity to power 111 homes. There are several wind energy projects in the pipeline. Pakistan has almost 1 GW of wind power projects under construction.[108] The next section will examine the major wind energy projects in progress and already completed.

a.   Fifty megawatt wind energy project in the Jhimpir Thatta District

On 24 December 2012 Pakistan's President, Asif Ali Zardari, inaugurated the country's first 50-MW wind power project which was to be operational in January 2013. The project was constructed at Jhimpir in the district of Thatta. It is only 144 kilometers away from Pakistan's largest city and commercial hub. The project was executed by FFC Energy Limited, a subsidiary of Fauji Fertilizer Company, Ltd. (FFCL), and Zorlu Energy Pakistan, the local subsidiary of a Turkish company. The project of 33 German-made wind turbines was funded by ADB and was completed

---

[108]   Haris Anwer, 'Pakistan Offers Renewable-Energy Incentives to Tackle Shortages,' Bloomberg, 21 Aug. 2011, http://www.bloomberg.com/news/2011-08-25/pakistan-offers-renewable-energy-incentives-to-tackle-shortages.html (last accessed 4 Jan. 2013), stating: 'Pakistan has almost 1 GW of wind-power projects under construction or with financing agreed upon and 498.5 megawatts more of plants announced, according to Bloomberg New Energy Finance data.'

within a year. According to the project director of the Fauji Fertiliser Company Energy Limited FFCEL Wind Farm, Brigadier Izaz, the total cost of the wind energy project was $134 million.[109]

The average wind speed at Jhimpir in the Thatta district, the site of the project, is 7.8 meters per second. The hot climate versions of the turbine Nordex S77/1500 series, specially designed for medium-strong wind conditions, are installed in this project. Nordex AG is producing these turbines at its facilities in China.

b.   Zorlu Enerji Electrik Uretim (ZEEU) project

Zorlu Enerji Electrik Uretim, a subsidiary of a Turkish company, and the AES Corporation, a US company, are working on other Pakistani wind energy projects. Zorlu Enerji Electrik Uretim (ZEEU) is involved in a wind energy upgrade project located in Jhimpir, 100 kilometers northeast of Karachi. ZEEU announced that it would use a $36.8 million loan to increase the output of its wind farm. The current capacity of only 6 MW will increase to a total of 56.4 MW after the installation of more turbines. The existing 6-MW wind farm project is selling power to the Hyderabad Electric Supply Company but, on the completion of its new phase, the 56.4-MW wind farm will supply power to the national grid through the National Transmission and Dispatch Company.

Pakistan also has entered into an agreement with the US AES Corporation.[110] AES agreed to build a wind farm to produce 150 MW of electricity near Karachi. The $375 million project is scheduled to be completed in two years at three sites in the Gharo Corridor in Thatta district of Sindh with assistance from the United States Agency for International Development (USAID).

L.   RENEWABLE ENERGY DILEMMA IN PAKISTAN

Renewable energy resources are widespread and abundant in Pakistan. The question arises as to why renewable energy development was virtually non-existent in the country. In my recent meeting (August 2012) the Prime Minister Raja Pervez Ashraf responded to the question in these words:

---

[109] 'President to Inaugurate Jhimpir Wind Farm Today,' ARY News, 24 December 2012. http://www.arynews.tv/english/newsdetail.asp?nid=68137. (last visited 5 Jan. 2013).
[110] AES owns and operates a diverse portfolio of electricity generation and distribution businesses. See AES, http://www.aes.com/aes/index?page=home (last visited 29 Dec. 2012).

Energy production is a long-term play and it demands the heavy bets and huge risks. The private and multinational investors' confidence is crucial in long-term energy infrastructure development. In the past our governments had a very short vision and failed to offer private investors economically feasible opportunities. Their policies could initiate only short-term import/export contracts. Our government is working very hard to provide a more stable and investor friendly environment for the private and international investors.[111]

Indeed there are numerous causes for slow growth in the renewable energy sector. We will examine below a few primary reasons.

1.   Lack of Funding, Finance, and Financial Incentives

The country has a very large potential for renewable energy resources but only a few projects have been completed. A lack of major financial incentives for solar panels and wind turbines prior to 2006 is partially to be blamed. India, Pakistan's neighbor right next door, although having less potential for wind, is the world's fourth largest producer of wind energy. India's installed wind turbines are capable of generating 7093 MW.[112] In India like other front runners (Germany, Spain, and some states of the United States) installed wind turbines and solar panels are guaranteed a certain rate per kilowatt-hour.

According to the Ministry of New and Renewable Energy, the government of India, in most areas, guarantees between 2500 and 4800 rupees for the installation of solar panels, and between 250,000 and 300,000 rupees are granted for the wind turbines installed.[113] The proper financial incentives can help to reduce the cost of energy as well. In Pakistan the cost of wind is 7 cents per kilowatt-hour, compared to India where the cost is only between 2 and 2.5 cents per kilowatt-hour.

Recently, a number of international organizations have played an increasingly important role in the renewable electricity generation projects.

---

[111]   The meeting took place in Prime Minister's House in Islamabad. Renewable energy was not on the agenda but the Prime Minister mentioned the energy issues. He also mentioned the International Renewable Energy Agency (IRENA) and his personal involvement in IRENA. He hoped that IRENA would not only provide guidance and assistance to Pakistan but to other developing countries as well.

[112]   In front of India are Germany at 21,283 MW, Spain at 13,400 MW, and the US at 12,934 MW.

[113]   GLOBE-Net, 'India: Renewable Energy Market,' http://www.globe-net.ca/market_reports/ index.cfm?ID_Report=1069 3. Also see Ministry of New and Renewable Energy Sources, 'CFA Provided under Various Renewable Energy Schemes/Programmes,' http://mnes.nic.in/cfa-schemes-programmes.htm. (last visited 5 Jan. 2013).

International financial institutions have also worked to support these projects. Moreover, several countries and multinational companies have been working with Pakistan to initiate energy from renewable sources.[114] The government of Pakistan now offers attractive tariffs (cost plus, with up to 17% ROE), indexed to inflation and exchange rate variation (rupee/dollar) as detailed above. Euro/dollar parity is allowed and issuance of corporate registered bonds is also allowed as described above.[115]

2. Lack of Technological Resources and Support

One obstacle to renewable energy progress is the country's insufficiency in the needed technologies. Pakistan is importing renewable technology from China and Germany. More research work is vital for the development of local production of solar panels, wind turbines, and other technologies to lower the cost of renewable energy.

All renewable power-generating projects require large numbers of skilled and technical workers. As these projects are new in the country, Pakistan lacks adequate labor and technical support. Without required technical knowledge, education, and training, locally produced renewable projects cannot be materialized.

3. Better Energy Management

Energy management is a vital issue in Pakistan. At present many departments of the government, including the Ministry of Water and Power, the Ministry of Oil and Gas, the Oil and Gas Regulatory Authority (OGRA), AEDB, NEPRA, PPIB, WAPDA, NTDC, the Sarhad Hydro Development Organization (SHYDO), and several more stakeholders, are directly or indirectly involved in energy production and management. Pakistan is one of those developing countries where there is poor coordination and cooperation among government departments. Involving so many departments is costly and creates ambiguity among departments and discourages new investors and producers. The current regime lacks an information sharing system among the various departments. Most of the efforts, especially seminars and conferences, prove to be overlapping and nonproductive. A more harmonious and effective energy management regime is vital to implement renewable energy policies in a more effective

---

[114] Ibid.
[115] See supra note 96. To view Alternative Renewable Energy Policy (ARE 2011), visit AEDB website, http://www.aedb.org. (last visited 4 Jan. 2013).

fashion. Department heads are appointed based on their seniority not on their knowledge and experience in the energy field.

M. CONCLUSION

Pakistan is an energy deficient country. The country's per capita energy consumption (3894 kWh) is less than the world average use (17620 kWh).[116] Pakistan is facing severe energy crises, and blackout periods are increasing day by day.[117] Power shortages are impacting the quality of life, which is leading to increased unemployment, poverty, and perilously escalating social unrest. The existing shortage of electricity and the constant rise in its use requires more electricity production. Close to half of the country's population lacks access to energy and they have a valid claim for their electrification.[118]

The global upsurge in the cost of fossil fuels is making it unaffordable to import these to Pakistan.[119] Several factors including fluctuations in oil price in international markets, capacity of aged equipment to generate the required electricity, and a poor distribution system are responsible for the energy shortage. Considering these factors and the fact that 70 percent of the population lives in a rural setting, the generation of localized energy seems to be a more practical option. The process of connecting remote areas with the main grid centers is costly and time-consuming.[120] The abundant availability of renewable resources in the country and the more

---

[116] Muhammad Shahid Khalil, Nasim A. Khan, and Irfan Afzal Mirza, 'Renewable Energy in Pakistan: Status and Trends, An Overview of Energy Equation of Pakistan,' http://202.83.164.28/mowp/userfiles1/file/uploads/publica tions/repk.pdf. (last accessed 5 Jan. 2013).

[117] Like many other Asian countries energy consumption in Pakistan is rising due to economic growth and net income increase. Changes in consumer lifestyle increased the use of luxury transportation, bigger houses, and electronic goods.

[118] In Pakistan only 55% of the population has energy access. In the province of Baluchistan approximately 90% of people live without electricity. See Khalil et al., supra note 116.

[119] Ibid. Pakistan is considered a net energy importer. Only approximately one sixth of the country's current oil requirements are produced locally.

[120] See Khalil et al., supra note 116, stating: 'These villages are separated by large distances with absolutely no approach roads. Transmission lines are very expensive in this area and there is no chance of grid connection in the near future. In remote areas, houses are mostly "kacha hut type" and light is their only requirement. Most of the houses consist of one room only. The electric requirement for each house varies from 50 watt to 100 watt maximum. Solar energy is the only and best solution for these areas.'

localized aspect of renewable energy offer a better choice for the much needed energy production. Renewable energy will help to improve the country's energy security, and a lower reliance on fossil fuels.

Recently the government has prioritized renewable energy production in the country. The government has made serious efforts to install and expand the use of solar and wind energy. A 50-MW wind energy farm has been inaugurated by the President and is expected to start full service in January 2013. Realistically it will be too soon to criticize these projects, especially the country's first, recently completed 50-MW wind energy plant and another under construction.

In the past, renewable energy resources remained virtually untapped in Pakistan. The study concludes that various factors are responsible for the slow uptake of renewable energy generation in Pakistan. Lack of involvement of the private sector and investment, insufficient funding and needed financing, deficiency of grants and financial incentives, lack of development of research work to produce the renewable energy, an untrustworthy investment environment, and political volatility are considered major impediments. Energy production is a long-term play and requires political stability besides favorable long-term investment policies for investors.

The study identified the growing challenges of energy in Pakistan and concludes that renewable energy projects seem to be the only way out. Pakistan's energy dilemma faces manifold obstacles but the government and Prime Minister Raja Pervez Ashraf (who has served as Minister of Water and Power) seem to understand the importance of renewable energy. He comprehends that much more needs to be done. His government is determined to push the policies to move towards the right direction.

The study concludes that the government needs to work on both short-term and long-term projects and policies. The short-term projects should include a plan for conservation and more efficient use of energy[121] because saving 1 MW is better environmentally and less expensive than generating 1 MW. The study concludes that more generous incentives and awareness are needed to encourage individuals to install solar home systems, solar water heaters, and solar water pumping. Local and city governments must use solar streetlights, and use renewable energy in the public places. It will

---

[121] Something is more energy efficient if it delivers more services for the same energy input, or the same services for less energy input. For example, when a compact florescent light (CFL) bulb uses less energy than an incandescent bulb to produce the same amount of light, the CFL is considered to be more energy efficient. See International Energy Agency, 'What is Energy Efficiency?', http://www.iea.org/index_info.asp?id=2401. (last visited 5 Jan. 2013).

not only help to curb the current shortage, it will also educate the citizens and will inspire them to use the renewable energy.

The study noted that biogas technologies needed to be improved and used as much as possible. Clean gas produced by biogas technology will reduce the reliance on and use of fossil fuel. It will also help to recycle agro-animal residues.

The study finds that there is no effective plan in place to encourage and accommodate individual needs. Involvement of citizens in this campaign is important. Incentives and financial support of individual consumers is a must. Presently there is no major local manufacturing of solar panels. If solar panels are manufactured in Pakistan, it will cost less and they will be easy to install and maintain by the local factory trained technicians. The government must subsidize renewable energy and provide financial support and easy financing for individual domestic users, which will release the pressure on the national grid.

Energy is one of the reasons foreign investment has not returned to Pakistan. Pakistan must address long standing issues related to government revenues and energy production in order to spur the amount of economic growth that will be necessary to employ its growing and rapidly urbanizing population, more than half of which is under 22. Energy production is necessary for the much needed economic growth in Pakistan.[122] In the long term the government of Pakistan must build up the confidence of national and international investors to invest in its renewable energy sector. Past incentives have done very little to stimulate growth in renewable energy. The government can achieve this goal by providing more generous incentives, including easy financing, and more attractive tariff and energy buy-back contracts to encourage renewable energy production on a commercial basis. The government must facilitate the private investors with suitable infrastructure and by acquiring the necessary land. No doubt renewable energy is a new and expensive technology. Its promotion will demand commitment and patronage from the government and more awareness by the consumer. It is not just energy; it is renewable and sustainable energy.

Besides the logical reasons there is a more valid demand to follow the renewable energy path. As the world wakes up to the reality of climate change and as a responsible member of the international community,

---

[122]   Central Intelligence Agency, The World Fact Book, supra note 26: 'Pakistan must address long standing issues related to government revenues and energy production in order to spur the amount of economic growth that will be necessary to employ its growing and rapidly urbanizing population, more than half of which is under 22.

Pakistan is at the frontline of climate change negotiations. Pakistan's electricity will increasingly have to come from renewable sources. Pakistan is in a good position to exploit available renewable resources for a greener energy future. Pakistan contributes only about 0.8 percent of global greenhouse gas emissions but its energy sector is responsible for half of this amount.

Societies across the globe are developing technologically to achieve a renewable energy transition. Pakistan also must carry out case studies of the projects in the countries that have made successful transitions to renewable energies. Studies of these success stories will overcome the challenges and help to repeat those models in Pakistan.

N. LESSONS LEARNED

The lessons to be learned from Pakistan's renewable energy initiatives are pretty well described in the sections on the renewable energy dilemma in Pakistan and the conclusion. As in most developing countries, there is a need for more funding and more expertise, technical resources, and training. There are serious problems of overlapping departmental responsibilities and clarity of authority. The involvement of the public in project development and implementation is vital. Corruption by the large interests involved in promoting renewable energy is a serious problem.

The emphasis on decentralized solar and wind energy for the large proportion of the population who live in rural areas is an important asset for the program. The emphasis on energy efficiency should be very helpful. The creation of an alternate energy development board to coordinate renewable energy activities is a particularly constructive innovation, but the board's authority, relationship with other government departments and with the local governments needs to be clarified. The collaboration with other countries and international and regional organizations also is very useful.

Pakistan, with great energy needs, seems to be off to a promising start on its renewable energy programs, particularly impressive in view of its financial difficulties and its many internal and international political challenges.

9. Conclusion

Richard L. Ottinger

From the case studies of renewable energy initiatives analysed above, from both least developed countries with poor economies and the richest emerging countries, it can be seen that renewable energy can be of great help in: strengthening their economies; creating local jobs; training workers, contractors, financial institutions and government officials in renewable energy technology economics and skills; relieving dependence on imports of expensive, unreliable, highly polluting fuels that often create great risks for human health and safety; improving energy security; reducing emissions of greenhouse gasses; and improving the welfare of women and children now dependent on gathering wood and burning it for heating and cooking at great risk and depriving them of educational opportunities.

The case study analyses also demonstrate that initiating renewable energy projects is a very complicated task which if done wrong can: be uneconomic; promote unsustainable exploitation of local resources for the prime benefit of project developers; cause great environmental damage including increasing greenhouse gas emissions; displace food crops; pollute air, land and water supplies; exploit and displace local labor with unlivable wages, dismal living conditions and risks to their health and safety; deprive local citizens of participation in the design and implementation of projects affecting their lives, including the rights to their property; enable conflicts between government agencies designated to implement environmental laws; and promote corruption that undermines the implementation of environmental laws.

The case studies show the importance of dedicated and knowledgeable leadership in: careful planning and execution of projects; public education and full disclosure of the costs and benefits of projects; participation for all citizens affected and officials responsible for the design and implementation of projects; obtaining adequate financing to support the projects; assuring that the contractors selected are qualified to perform the projects and do so competently, taking responsibility for maintenance of equipment and supplying any spare parts that may be needed; compensating anyone damaged or adversely affected economically by their work and providing in advance insurance or funds for such compensation; providing for regular monitoring and evaluation of projects after construction with full liability for correcting defects; requiring that developers conform to all applicable labor and environmental laws with heavy penalties and adequate enforcement personnel to enforce these laws; and assuring that contracts with developers provide a fair share of the project profits to the countries involved.

International agencies, project funders, and purchasing countries are just beginning to become aware of and demand compliance with these requirements for successful and sustainable projects. The best way to assure that such requirements are met is to provide standards, criteria and certification before an investor will provide funds for a renewable energy project. One of the best ways to assure compliance with regulatory requirements and avoid corruption is to provide for a certificate of compliance from a qualified independent agency and to persuade importing countries to purchase only certified products. This is being done by the Roundtables for Sustainable Biofuels and Sustainable Palm Oil with oversight by NGOs like the World Resources Institute.

The temptations of short-term profits from renewable energy should not persuade countries to engage in short-cuts for approval of projects that do not meet these basic sustainability requirements. To do so can result in serious environmental damage, exploitation of resources and labor, and the economic failure of projects.

Index

separated plutonium 50
service stations, biogas 14–15, 19
Shanghai
  Donghai Bridge project 33–8
  Jiao Tong Energy Center 30
  Jiao Tong University Institute 27
  Municipal Economic Commission
    27
  Solar Energy 20
  'Sunshine Fund' 29, 30, 31
Shramik Bandhu 100
Sinovel 34, 35–6
slurry 89–90, 97–8, 99
smart grids 38
SNV 187
'Social Fuel Seal' program 132, 160
social justice (China) 3
social problems 129, 133, 145–7
social protection (India) 81–2
soil 73–4, 150
solar combined pilot plant (Aïn Beni
  Mathar) 68–80
Solar Electric Generating Systems 72
solar energy
  Brazil 108, 112, 122–4, 131–2
  China 25–33
  Morocco 68–80
  Pakistan 175, 181–2, 188, 190,
    191–2, 194–5, 197–9
  Philippines 58–9, 61, 63–4, 65, 66
solar photovoltaic energy 25–33, 65
solar roof generation 27–33
solar thermal utilization (China)
  19–25
Solar Valley project (China) 20–22
solar water heaters 19–20, 22–5, 197
State Bank of India 86
State Nuclear Power Technology
  Corporation (China) 45–6, 51
straw stalk biogas program 10, 11
subsidies
  Brazil 113, 117, 120–21, 126–7,
    130–31, 132–3
  China 7, 8, 9–10, 13–14, 16, 18–19,
    25, 28–31, 33, 36–7
  India 90–91, 93–5, 99, 103, 105
  Indonesia 139–41, 144–5, 152, 158,
    169, 170
  Morocco 69, 75
  Philippines 62, 67

sugar cane production
  Brazil 107, 109, 124–7, 132–4
  India 87–8
Suharto, President 138, 145–6
Sunech Power 20
supply chains 55, 144
sustainability 198, 201
  India 82–3, 89, 96, 101
  Indonesia 133, 137, 143–4, 153, 157,
    159–62, 171–2

Taishan nuclear reactors 46, 57
Tata Energy Research Institute 91
tax incentives 188
  Brazil 112, 115, 128
  Indonesia 140, 159–60
  Philippines 62, 67
technology
  biogas (India) 91, 96, 103–4
  nuclear (China) 46–7, 49–50
  transfer 58, 78, 99, 101, 183
Telesp 128
Thardeep Rural Development
  Programme 192
Three Gorges Dam (China) 39–40
Tianwen Group 20
'Township Electrification Program' 26
training
  Brazil 117–18, 131, 132
  China 9, 10, 32, 51–2
  India 82, 93–100, 102–4
  Indonesia 145–7, 159
  Morocco 73, 77, 80
transesterification plants 86, 87
Transmigration schemes 144, 146–7
transparency 78, 131
  Indonesia 144, 155–6, 158–9, 161,
    165–6, 170
transport 180
  Brazil 124–9, 132–4
  India 83, 85
  Indonesia 137, 141, 143
Trina Solar 21
Turbo Institute of Slovenia 183
turn-key job fee 92, 93, 94

UN–REDD 164
underground research laboratory
  49–50
Unilever 162, 172